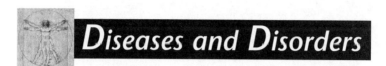

Diseases and Disorders

Teen Depression

by Michael J. Martin

LUCENT BOOKS
An imprint of Thomson Gale, a part of The Thomson Corporation

Detroit • New York • San Francisco • San Diego • New Haven, Conn.
Waterville, Maine • London • Munich

© 2005 Thomson Gale, a part of The Thomson Corporation.

Thomson and Star Logo are trademarks and Gale and Lucent Books are registered trademarks used herein under license.

For more information, contact
Lucent Books
27500 Drake Rd.
Farmington Hills, MI 48331-3535
Or you can visit our Internet site at http://www.gale.com

ALL RIGHTS RESERVED.
No part of this work covered by the copyright hereon may be reproduced or used in any form or by any means—graphic, electronic, or mechanical, including photocopying, recording, taping, Web distribution, or information storage retrieval systems—without the written permission of the publisher.

Every effort has been made to trace the owners of copyrighted material.

LIBRARY OF CONGRESS CATALOGING-IN-PUBLICATION DATA

Martin, Michael J.
  Teen Depression / by Michael J. Martin
    p. cm. — (Overview series)
  Includes bibliographical references and index.
  Summary: Describes teen depression, its causes, and how it can be treated and prevented.
    ISBN: 1-59018-502-1

Printed in the United States of America

Table of Contents

Foreword	4
Introduction The Hopelessness Disease	6
Chapter 1 More than Sadness	10
Chapter 2 Down in the Dumps: Why Teens Get Depressed	20
Chapter 3 Helpless and Hopeless: The Signs of Depression	33
Chapter 4 When Depression Turns Deadly	47
Chapter 5 Restoring Hope: Treating Depression	57
Chapter 6 Keeping the Blues at Bay	70
Notes	80
Organizations to Contact	84
For Further Reading	86
Works Consulted	88
Index	91
Picture Credits	95
About the Author	96

Foreword

"The Most Difficult Puzzles Ever Devised"

CHARLES BEST, ONE of the pioneers in the search for a cure for diabetes, once explained what it is about medical research that intrigued him so. "It's not just the gratification of knowing one is helping people," he confided, "although that probably is a more heroic and selfless motivation. Those feelings may enter in, but truly, what I find best is the feeling of going toe to toe with nature, of trying to solve the most difficult puzzles ever devised. The answers are there somewhere, those keys that will solve the puzzle and make the patient well. But how will those keys be found?"

Since the dawn of civilization, nothing has so puzzled people—and often frightened them, as well—as the onset of illness in a body or mind that had seemed healthy before. A seizure, the inability of a heart to pump, the sudden deterioration of muscle tone in a small child—being unable to reverse such conditions or even to understand why they occur was unspeakably frustrating to healers. Even before there were names for such conditions, even before they were understood at all, each was a reminder of how complex the human body was, and how vulnerable.

While our grappling with understanding diseases has been frustrating at times, it has also provided some of humankind's most heroic accomplishments. Alexander Fleming's accidental discovery in 1928 of a mold that could be turned into penicillin

Foreword

has resulted in the saving of untold millions of lives. The isolation of the enzyme insulin has reversed what was once a death sentence for anyone with diabetes. There have been great strides in combating conditions for which there is not yet a cure, too. Medicines can help AIDS patients live longer, diagnostic tools such as mammography and ultrasounds can help doctors find tumors while they are treatable, and laser surgery techniques have made the most intricate, minute operations routine.

This "toe-to-toe" competition with diseases and disorders is even more remarkable when seen in a historical continuum. An astonishing amount of progress has been made in a very short time. Just two hundred years ago, the existence of germs as a cause of some diseases was unknown. In fact, it was less than 150 years ago that a British surgeon named Joseph Lister had difficulty persuading his fellow doctors that washing their hands before delivering a baby might increase the chances of a healthy delivery (especially if they had just attended to a diseased patient)!

Each book in Lucent's Diseases and Disorders series explores a disease or disorder and the knowledge that has been accumulated (or discarded) by doctors through the years. Each book also examines the tools used for pinpointing a diagnosis, as well as the various means that are used to treat or cure a disease. Finally, new ideas are presented—techniques or medicines that may be on the horizon.

Frustration and disappointment are still part of medicine, for not every disease or condition can be cured or prevented. But the limitations of knowledge are being pushed outward constantly; the "most difficult puzzles ever devised" are finding challengers every day.

Introduction

The Hopelessness Disease

WHEN JESSE, A first grader, threatened to shoot himself, no one took him seriously. Ten years later, when Jesse was a high school sophomore, neither his parents nor his friends suspected that depression was stalking him. After all, he was a young man who seemed to have everything going for him. Near the top of his class academically, Jesse played violin and guitar and was a starting pitcher on the baseball team.

Then, one day during a classroom discussion of *Catcher in the Rye* in English class, Jesse shocked his teacher and classmates by bursting into tears. His worried parents quickly scheduled a meeting with a psychiatrist. What they heard Jesse say at that first session terrified them: "The world is pushing me to do the right thing, but nothing changes. I just want to feel better, but I can't. If someone handed me a gun and loaded the gun and took the safety off, I'd stick it in my mouth. If someone gave me an ax, I'd chop my head off."[1]

Jesse is not alone in his anguish—or in his susceptibility to depression. A shocking number of adolescents share his feelings of helplessness and hopelessness. According to the U.S. Surgeon General's report, 3.5 million children and teenagers suffer from depression. And depression—especially serious depression—is a life-threatening condition. Each year, 3 million teens think about committing suicide. About four hundred thousand of them actually try to kill themselves.

It is a good bet that virtually every teen who thinks of suicide would agree wholeheartedly with Jesse that "I just want to feel

The Hopelessness Disease

better, but I can't." That is the uniquely painful and insidious thing about depression. It produces a curious kind of inertia that makes the solution of one's problems seem forever beyond reach. Paul C. Quinnett, author of *Suicide: The Forever Decision*, describes the hopelessness of severe depression this way: "It is the one human experience that, when it won't go away, makes us sicker and sicker until, when we are way down in the bottom of that black hole, we can't imagine ever feeling any better. I once asked a very depressed young man if he was anxious about something. 'No,' he sighed, 'I'm too depressed to be anxious.'"[2]

If someone were dying of terminal cancer and learned of a cure that could be had thousands of miles away, that person would probably hop on the first plane to try to save his or her life. Severely depressed people, on the other hand, "know" that they will never get better. Dick Cavett, the former television talk show host,

Depression is a serious disease that afflicts millions of people around the world, including large numbers of teens and children.

recalls the worst part about his struggle with severe depression this way: "What's really diabolical about it is that if there were a pill over there, 10 feet from me, that you could guarantee would lift me out of it, it would be too much trouble to go get it."[3]

Cavett is not a teenager, but his inability to help himself would sound familiar to millions of depressed teens. In some ways Cavett was luckier than many adolescent depressives because he knew what he was suffering from. Depressed teens frequently have no idea that they are depressed, and because their behavior is not always sad and lethargic—as it so often is with depressed adults—their parents often do not realize it either. Here is how a fifteen-year-old New York girl describes her behavior before she realized she was depressed: "You seem to get angry at everyone for no apparent reason. You always feel like crying. You have low self-esteem and you do very, very stupid things, very crazy things. Three times I attempted suicide. I still have one of the scars. It's permanent."[4]

People sometimes mistakenly believe that depression comes from weakness or some kind of character flaw. They may criticize the depressed person for acting lazy or not trying. They do not realize that depression is not an attitude one can change or a mood one can shake. When a person is depressed, all other feelings fade into pointlessness. Food does not taste good, no activity seems worth doing, and nothing seems funny. It is as if the teen has entered a gray world where all the things that make life worth living have lost their value.

Knowledge Is Power

Yet, as with any difficulty, the crucial first step is recognizing exactly what one is up against. As the American Psychiatric Association says, "one of the biggest problems is that people do not understand depression and even deny its existence."[5] Yet, when people begin to learn the true nature of depression, everything changes. Once teens like Jesse accept that they can get better, they can begin climbing out of the black hole of depression.

Unfortunately, the National Institute of Mental Health says that less than one-third of all Americans who become depressed will ever get treatment. That is unfortunate, since there are millions of teens who, given help, could triumph over depression the way that Cait Irwin did. At age thirteen Cait struggled with a depression so severe that she contemplated suicide. She came to think of her affliction as a "beast." Eventually, that beast caused her parents to hospitalize her for her own safety. After successful treatment for her depression, a healthy and happy Cait returned to school, went on to college, and even started her own business. Instead of regretting what happened to her, Cait considers it empowering: "The 'beauty' of depression is there's much more to gain than lose from the experience. The feeling you'll get when you face and defeat the beast is worth the fight. When your battle is won you'll be able to face anything on the horizon."[6]

Chapter 1

More than Sadness

SIXTEEN-YEAR-OLD Courtney, a straight-A student and varsity athlete, was crying uncontrollably. Alone in her room at home, she felt the most intense emotional pain of her entire life, yet she had no idea what was causing it. "That's the worst part about it—crying and feeling like your world is ending but not knowing what is causing you to ache,"[7] Courtney recalled many months later.

Courtney's friends could not help her either. They had no idea what was going on inside her head because "on the outside I was a regular high school girl, but on the inside I was falling apart. I was too embarrassed to tell them: 'I cry and feel sad all the time and I don't know why.'"[8]

Depressions like the one Courtney experienced are far more common and far more serious than most people realize. Depression is a potentially devastating disease with far-reaching implications for affected teens, their families, and communities. Yet, with proper diagnosis and treatment the situation is far from hopeless.

Greatest Cause of Disability

Courtney received treatment that helped her get well. But experiences like hers are repeated daily in almost every community in America. Appearing without warning or explanation, depression can be particularly harmful for teens. Not only do adolescents with depression suffer during a crucial stage of their physical and mental development, they remain at increased risk for further episodes of depression for the rest of their lives.

According to the World Health Organization, depression is the world's greatest cause of disability and is responsible for more human misery than any other single disease. Even though it is considered a mental disorder, depression's effects can be more disabling than many physical disorders. Being unable to eat or sleep properly or being constantly fatigued—all typical signs of depression—can make life just as difficult as dealing with disorders like lung disease, arthritis, and diabetes.

Ignored and Underestimated

Recent statistics underscore how widespread teen depression has become in this country. Somewhere between 10 and 15 percent of all teens suffer from depression at any one time. But there are undoubtedly millions of cases of teen depression that never get reported. A study done by the University of Oregon found that one out of every five teens had a major episode of depression that went untreated during their adolescence. In fact, most depressed teens do not receive the help they need.

Two teenage girls discuss their problems. Unfortunately, most depressed teens are too embarrassed by their condition to turn to friends for help.

One reason that the problem of depression is not regarded as seriously as it might be is confusion about the meaning of the word. Teens will often claim they are depressed about something when what they really mean is that they are upset or unhappy. It is not unusual to hear an adolescent say something like, "I'm really depressed about my grade on the history test" or "I'm depressed about how my friends treat me." Getting a poor grade on a test or suddenly being ignored by friends is sufficient reason to feel sad or upset, but it is not true depression.

More than Mood Swings

All teens, as do all adults, go through periods when they feel down. But those kinds of mood swings are normal—and they happen more often during adolescence than at any other time of life. True depression, however, does not go away after a few hours or days. Weeks, months, and even years can pass before a depressed teen begins to feel better. It is a serious mental disorder—some have called it a malignant sadness—that can have long-term effects on one's mood and behavior. After a while, the teen can no longer function normally. The mood completely takes over his or her life.

Depression is not just a phase that the teen is going through. Nor is it being sad, unhappy, or even grief stricken. It is a mental disorder that should not be ignored in the hope that it will go away. Doctors define depression as a condition where marked changes in one's mood and other behavior have lasted at least two weeks. It is accompanied by a whole cluster of related symptoms—a "syndrome," in medical terms. The syndrome for depression includes appetite problems; lack of energy; feelings of worthlessness, hopelessness, or guilt; problems in concentration; sleep problems; and in thinking about or attempting suicide. If at least four of these six problems occur almost every day (suicidal thoughts do not have to occur daily to make the list) then a major depression is suspected.

The changes in mood and behavior brought on by prolonged depression are accompanied by subtle changes in the brain. Whether depression causes those changes or is a result of them is not known. But changes in the way the brain works can lead to numerous problems—both in one's moods and physical health.

Two of the more common symptoms of depression are a lack of focus and an inability to concentrate for extended periods of time.

In a depressed person the brain and body do not always communicate well with each other. A severely depressed person, for example, may not eat or sleep when he or she should.

But diagnosing the disease can be difficult. Not every depressed teen has the same symptoms. One depressed teen might be angry all the time, another might be hyperactive, while another might refuse to leave his or her room. All these behaviors are possible with depression, but they could also be signs of some other problem. Furthermore, teen depression can look completely different than adult depression, so it often takes an expert to make the right diagnosis.

It Is Not a Choice

Unfortunately, because of continued ignorance about the serious nature of depression, shame and guilt are too often associated with becoming depressed. Parents or friends will sometimes tell a depressed teen to "Snap out of it!" as if the depression were a choice the teen were making. It is not. When Courtney's sisters became concerned about her depression and asked what was wrong, her reply was typical: "I don't know what's wrong! I don't like myself and I don't want to do anything or see anyone and I feel sad and confused and alone and helpless and I do want to snap out of this, I do! But I can't!"[9]

When her family showed their concern for her, Courtney felt guilty about causing them to worry. She was also ashamed of feeling so bad, since she could think of no reason for her mood. Maybe, she thought, she was just weak and selfish. Such thoughts further deepened her depression and led to suicidal thoughts that maybe her family would be better off without her.

Parental concern over teen depression can often make teenagers feel guilt for causing their parents to worry.

"'I just don't think I can get through this' ran through my head every minute of every day," Courtney recalls. "Whenever I hear people say that 'suicide is the most selfish act of any person,' I just think that they have no concept of what it's like to be depressed."[10]

Doing Battle with a Beast

Cait Irwin knows exactly what Courtney was feeling. After recovering from her depression, she wrote the book *Conquering the Beast Within* in order to help others to get through the disease. She compares depression to a monstrous beast trying to take over one's mind. "The most important thing to remember about depression and the beast is that . . . you and the beast are not the same, you are two separate beings," she says. "It's perfectly clear it's not your fault nor is it a character flaw. You are you! And the beast is depression."[11]

Instead of something to be ashamed of, Cait tells teens that they should think of depression as a curable illness, not a personality weakness. No one, she points out, feels guilty for being treated for a broken leg. She argues that, since the causes of depression are outside one's control, one should feel the same lack of shame about being treated for depression.

Cait Irwin's message of tolerance and hope is the kind of positive news that teens suffering from depression most need to hear. Like Courtney alone in her room, teen depressives live in a bleak world. They are overwhelmed by feelings of despair and self-hatred, as well as the physical symptoms that can accompany deep despair. They often have sleeping and eating problems along with bodily aches and pains that make day-to-day life an unending ordeal.

Help Not Wanted

The one feeling that is nearly always present in depression—and is perhaps the most painful feeling of all—is utter hopelessness. Depressed teens tend to dread the future and feel they are doomed to failure. "Occasionally a teacher or counselor would express concern and offer help," says Sylvia of her teen depression. "But at that point, I couldn't see any hope and shrugged off

Rock star Kurt Cobain began to suffer from depression as a teenager. After refusing to seek treatment, he committed suicide in 1994.

their offers of assistance. Somewhere inside me, I did want help, but I just couldn't reach for it."[12]

Maryann, a depressed teen who attempted suicide three times, recalls being unable to experience happiness. "When adolescent depression attacks, it totally destroys you," she says. "It gives you no hope, no dreams, nothing."[13] Because of their self-degrading thoughts, depressed teens often refuse help and deny that things could ever get better. They feel so worthless that acceptance of any kind of help seems pointless. Kurt Cobain, the rock star who was the lead singer of the band Nirvana before his suicide, was very depressed as a teen. That depression was never treated and made it impossible for him to enjoy his success in music. Even after he became a star, he still wrote lines like "I hate myself most of all,"[14] in his journal.

Cobain's friends tried to get him to do something about his depression, but he rejected all offers of help. Yet his journals indicate that a part of him desperately wanted the situation to change. "I'm so tired of crying and dreaming, I'm soo soo alone," says one entry. "Isn't there anyone out there? Please help me. HELP ME!"[15]

The Cost of Feeling Blue

Cobain's suicide in 1994 was both a loss to the world of music and an extreme example of where depression can lead. It is unfortunate that he never got the help he needed, yet his case is far from unusual. Depression is severely underdiagnosed. Parents and doctors do not always recognize the signs of depression and teens often do not have any idea why they feel so bad. The underdiagnosis of the disease has serious consequences. Depressed teens tend to either isolate themselves from their families or become angry and irritable. In either case, family relationships suffer. Since depressed teens do not have the energy to devote to relationships with friends, their social development is stunted. And, no matter what kind of student a teen is, his or her grades are likely to drop when they become truly depressed. Because they are so unhappy, depressed teens tend to look for relief in sex, drugs, or alcohol abuse. Such relief is only temporary and often leads to even greater problems. Depressed girls, for example, are at a much higher risk for unwanted pregnancies.

The failure of these depression-motivated strategies is undoubtedly a major factor in the four hundred thousand or so suicide attempts made by teens each year. That averages out to well over one thousand suicide attempts per day—every single day of the year. Most, if not all teens who attempt suicide, are suffering from depression. Early diagnosis and treatment of their disorder could save countless lives.

The Number One Mental Illness

Despite all this misery, up until a couple of decades ago it was not thought that teens could even get depression. According to psychiatrist Harold S. Koplewicz, director of the New York University Child Center, it was believed that "young people had neither sufficiently formed egos, nor the brain development to cause the kind of chemical imbalance that is at the root of clinical depression."[16] That belief is now known to be untrue—even babies can show signs of depression. Meanwhile, depression has become the most common mental illness among teens. In any given year, about 8 percent of the adolescent population will show signs of major depression. That compares to about 5 percent for adults.

Triggers, but No Smoking Gun

Although teens seem more prone to depression than adults, no one really knows why that is so. Since a person is more likely to get depression if someone else in the family has had it, heredity has always been suspected. Perhaps something happens at puberty that triggers a hereditary propensity for depression. Another explanation is that the increased hormone levels of teens can create some sort of imbalance within the brain that leads to depression. The increased social and physical stresses of adolescence have also been cited as possible causes.

Recent research suggests that adolescent brains seem to be undergoing extensive changes during the same period—ages fourteen to seventeen—when depression is most common. Perhaps that is more than coincidence.

The social stresses of interacting with peers of both sexes are a common trigger of depression in teens.

The Good News

The recent studies on the development of the adolescent brain are a sign of the growing awareness that teen depression is a serious problem—a problem that can affect a teen's future or even be life threatening. Yet, as big as the problem is, there are also ample reasons for hope. In almost all cases, the pain of depression does respond to treatment. And, depression is a time-limited disease. After six to eight months most teens do get better. Many teens, in fact, find that a diagnosis of depression is a huge relief. Contrary to what they thought, they are not going insane or dying of some terminal disease. Sometimes, giving a name to one's problem is the first step in overcoming it.

There is also reason for optimism regarding the treatment of depression. Doctors have more and more options. New kinds of short-term therapy seem to be beneficial for depressed teens. Drugs developed in the last few years have proved just as effective and are even safer than medication used in the past. And, no matter how depression is treated—with drugs, therapy, or a combination of both—early diagnosis reduces the risk of further episodes later in life.

No More Black Clouds

Caroline, age fourteen, was one of the lucky teens who got treatment before her depression got worse. She was so down that she did not have enough energy to answer the phone, open a schoolbook, or even put on her makeup in the morning. All she could do was sleep, but no matter how many hours she spent in bed, she still woke up tired and groggy.

Caroline's parents took her to see a therapist. When Caroline mentioned how hard it was to get out of bed in the morning, her therapist thought medication might be a good idea. "At first I couldn't feel anything," recalls Caroline, "but after a month the black cloud sort of lifted. Even my friends noticed I was in a better mood. Finally I could start taking care of myself like I used to. Like, you know, getting up in the morning, getting ready for school. It was a huge relief."[17]

Chapter 2

Down in the Dumps: Why Teens Get Depressed

PARENTS SOMETIMES EXPRESS amazement that adolescents can suffer from depression. Mindful of the daily responsibilities adults face, they will ask "What in the world could teens have to be depressed about?"

Margaret recalls that her condition worsened when her family could not accept that she was suffering from depression.

> When I was a teenager, I would say to my mother, "I feel depressed." She would scold me, "No you don't! You have nothing to feel depressed about!" But I did. I wasn't allowed to have any feelings. So when I was sad or angry or lonely I was told I wasn't. I began to think I was a bad girl to have those feelings. I learned to shut whole parts of me down. . . . I think depressed people have lost a sense of who they are.[18]

Adults may not be aware that the adolescents of today are subjected to a wide assortment of stresses—perhaps more than at any time in history. The nature of modern society itself makes it difficult for teens to discover who they are. That uncertainty undoubtedly contributes to the spread of depression. Back when America was largely rural, teens had jobs or chores that were vital for the family's well-being. Kids on farms, for example, might help out in the fields at harvest time. City kids might contribute much-needed income to the family by mowing lawns or running a paper route. Those kinds of activities were not always enjoyable, but they were important—and, regardless of age, people who feel needed and useful are rarely depressed.

Students and Shoppers

In today's affluent, technological society, few teens contribute to their family's economic survival. Adolescents are rarely needed to help out in the ways they once did. Some experts feel that lack of a vital role may have contributed to the rise of depression. Richard MacKenzie, director of the Division of Adolescent Medicine at Children's Hospital in Los Angeles, puts it bluntly. There is," he says, "no use for a teenager in American society today."[19]

Instead of working, young adults are expected to stay in school for years, often well into their twenties, so that they can qualify for a decent job. Better students face constant pressure to do well and it can seem as if the end of schooling is so far off that it will never come. That thought alone can lead to depression.

Meanwhile, advertisers relentlessly target adolescents. The modern American culture tends to see teens as consumers rather than young people with the potential to make valuable contributions to the world. They are encouraged to buy the hottest clothes, the latest music, and the trendiest items. Yet some teens discover

Young girls admire their reflections in a clothing store. The pressure to own the latest fashions and other trendy things leads some teens into depression.

that, no matter how much cool stuff they acquire, their possessions do not make them happy for very long. Ironically, guilt over not being happy about all that one possesses can lead sensitive teens deeper into guilt and depression.

Courtney felt that she had no right to feel as awful as she did: "After all, what the hell was I whining about? I had a loving family and wonderful friends. I was a straight-A student and a varsity athlete and involved in virtually every club and extracurricular [activity]. I was just weak and selfish. There were starving children in Ethiopia . . . they had a right to cry. Why the hell was I such a selfish wimp?"[20]

Searching for an Identity

Since the world around them so often marginalizes young people, they turn elsewhere to define themselves. They make strong commitments to their peers, to fashions, gangs, sports, and music. Unfortunately, those kinds of commitments are likely to increase confrontations with parents and teachers, who have trouble understanding what is so important about owning the latest hit CD or wearing a certain kind of shoe.

Anxiety and a persistent feeling of isolation from their peers make many teens highly susceptible to depression.

As teens work to find their place in the world, they may experience setbacks and disappointment. When a teen loses a boyfriend or girlfriend, fails to make a sports team, or sees a loved one die, he or she can be plunged into despair. Since most teenagers have limited exposure to emotional pain, drug or alcohol abuse can be particularly tempting. Experts like Dr. Gerald Oster, co-author of *Helping Your Depressed Teenager*, have noted a dramatic rise in teens' use of drugs and alcohol in the past twenty-five years, in correlation with the changing roles for young people in society. Drugs and alcohol can complicate one's adjustment to the frustration, disappointment, and loss that are an inevitable part of growing up.

Anxiety and Other Problems

Sixteen-year-old Andy was a good student who felt shy and anxious in school and did not consider himself part of the popular crowd. After struggling to ask a girl to go to a big dance with him and being turned down, he became totally discouraged. He felt certain his life would never change and that he would never have any real friends. Suicide seemed like the only way to relieve a pain he thought would never end. Hospitalized with other teens like himself after taking an overdose of pills, Andy was surprised to learn that other teens were just as unrealistically negative toward themselves as he had been. That helped him put things in perspective. Eventually he returned to school and began having more positive relationships with his peers.

Andy's anxiousness put him at risk for depression, but there are many other personal factors that can indicate a teen has an increased risk of depression. A history of physical abuse is one example. For obvious reasons, a teen who has been abused by a parent or guardian has good reason to feel depressed. And the abuse does not necessarily have to be physical. Emotional maltreatment or cruelty can be just as painful and degrading.

Family Factors

If a teen has relatives with a history of depression, he or she is more likely to suffer from depression, too. There is no firm proof that depression is a genetic disorder, but the evidence suggests that genetics does play a role. The mothers of depressed teenagers, for example, have high rates of depression. It has been estimated that between one-half and three-quarters of depressed kids have mothers who suffer from depression. This "family connection" is most evident in cases of depression that occur before the age of sixteen. After the age of sixteen, the risk of depression among teens seems to be roughly the same, regardless of whether or not one of the parents has had the disease.

Teens who have spent time in foster care are at a higher risk for depression, as are children who have attention problems in school. But risk never means that depression is inevitable. It does, however, seem to make a teen more vulnerable to the kinds of stresses that can trigger depression. The peak ages for the onset of depression are thirteen and fourteen, and a family crisis of some sort can bring on a bout of depression.

Before Matt's parents separated and his father developed colon cancer, he had been a happy, well-adjusted kid. Suddenly his safe, secure world was collapsing around him. It even seemed possible that one of his parents would soon die. Matt had trouble dealing with the changes emotionally. Two months later he was hospitalized with depression.

If parents have problems getting along, their children will have a higher risk of depression. Like Matt, Jeff was a well-adjusted, happy adolescent before his parents separated. He was in the tenth grade when it happened:

> After my dad left, my schoolwork started getting worse. I was sleeping a lot, and I was always in a bad mood. Nothing made me happy, not even baseball, which I love. . . . I think the worst part about being depressed was that I didn't enjoy anything. Nothing seemed like fun; nothing seemed interesting, no matter what I did. I knew something was wrong because I'm not a gloomy kid. But when my parents split up, everything got really bad. I just didn't feel my normal self. Now I feel a lot better. I'm not always gloomy.[21]

When parents separate or divorce it can be just as devastating as a death to their children—particularly if the missing parent comes into and out of the child's life intermittently. That can lead to a cycle of hope, disappointment, rejection, and despair. But even if the divorce is handled well, the children will likely face new kinds of stresses. Family finances can become strained because there are two households to pay for instead of one. Divorced parents are also at risk of becoming depressed, with little energy to spare for children who may be on the verge of depression themselves.

Remarriage of one or both of an adolescent's parents can also lead to an emotional crisis. Children who live in stepfamilies are more likely to admit to being lonely and depressed. One explanation is that remarried parents are so focused on making their new marriage work, they invest less time in their kids than single parents and parents in intact families.

When Families Do Not Function

Even when an adolescent's parents remain together, there are other kinds of stresses that can trigger depression. Many families are *dysfunctional*, a word that simply means they do not work well. There may be physical, emotional, or sexual abuse—or just an inabilty to communicate with one another. For whatever reason, the family's typical ways of relating to one another are not the best for a teenager's needs.

Teens who feel emotionally distant from their parents have an extremely high incidence of depression.

Steven Katz of Northwestern University Memorial Hospital studied how family interactions related to the onset of teenage depression. He discovered that "parents of depressed adolescents are more hostile and critical and more controlling toward their adolescents as well as less responsible in general toward their depressed adolescents than are parents of normal adolescents."[22]

Neglect or disinterest can be particularly devastating for teens. French researchers have shown that, while teens sometimes become depressed when their parents are overprotective, they feel most depressed when their parents show a lack of interest in them.

A related study done in France found that the most depressed teens were those who were given too much independence too soon. Instead of feeling empowered, they tended to feel rejected and unloved. Another kind of family dysfunction occurs when one or both parents are depressed themselves. As mentioned above, depressed parents can be emotionally unavailable when teens need help the most.

Grades a Source of Friction

Some parents, conversely, are overly involved with their children. Unrealistically high expectations can put great pressure on teens. Rick, a high school junior with a slight learning disability, was depressed and down on himself because he could not be what his parents wanted him to be:

> My older brother was this super athlete and top student and they think he's perfect. My grades . . . I mean, I think I work much harder in school than he did, but I don't get any credit for that. They just punish me for getting Cs and they think if I get a B that's no big deal, except it is for me. They expect me to be like my brother and I can't be—and they don't think I'm any good if I can't measure up. I feel like nothing I ever do will please them or make them proud. I wonder if they really love me. And I feel pretty much alone.[23]

Even when a teen meets a parent's high expectations, that does not necessarily mean all is well. Melissa, a bright seventeen-year-old planning on becoming a doctor, is upset because her parents take her success for granted. "Sometimes I feel like messing up or

Anxiety over academic performance can damage a teen's sense of self-worth and contribute to the onset of depression.

running away just to shake them up," she says. "I really get depressed when I feel like I'm their little 'success object.' I don't they even know or care about the person I am."[24]

While expectations about grades can be a major stress factor for the onset of depression, social relations at school may be even more critical. Adolescence is a time when children test their independence and begin moving away from their parents. At the same time, being accepted by one's peers assumes great importance. Studies have shown that negative evaluations by one's peers can severely damage a teen's fragile self-esteem. Kids who perceive themselves as "losers" are much more likely to suffer from depression. Rejection or harassment at school—and the social isolation that is likely to follow—can have devastating consequences. A Dutch study conducted by a Duke University researcher found that adolescents who have been rejected by their peers have a lifelong tendency toward depression.

Depression and Girls

Surprisingly, just being a girl is also a risk factor for depression. Girls who are sixteen or older are twice as likely as boys to suffer from depression. The reasons probably have to do with stresses that are unique to girls and make it harder for them to "fit in." Like boys, girls who do not do well in school feel low self-esteem. But girls who are good students also feel low self-esteem if they do not get positive feedback from their classmates. That kind of feedback, however, tends to go to girls who are attractive, regardless of how competent or smart they are.

As noted previously, divorce is a risk factor for depression. For unknown reasons, adolescent girls are three times more likely to have parents that are divorced than adolescent boys. "The weight of the evidence suggests that parents have a preference for boys,"[25] says Enrico Moretti, an economist at the University of California at Los Angeles who looked at the statistics. Regardless of the reason, more girls than boys must deal with school and divorce at the same time.

Girls are also more likely to be sexually harassed at school and the harassment can come at the worst possible moment for their self-esteem. For girls, puberty often coincides with the transition to middle school—an already stressful time for many young ado-

Girls are more likely than boys to be sexually harassed during adolescence, a situation that can compromise their self-esteem.

lescents. Boys, who start puberty later than girls, usually have adjusted to middle school before their bodies begin to change. That may be one reason why high school boys tend to have a much more positive self-image than high school girls do. Even worse, girls who come from a home where there is abuse or family conflicts are much more likely to begin puberty earlier.

Although teens of both sexes can be under intense social pressures and tensions, Harold S. Koplewicz, an expert in teenage depression, believes that girls are more vulnerable when something goes wrong. "They also cope differently," he adds. "Girls are thought to dwell more on broken relationships, to become more distressed when they are rejected, and that could make it easier for them to slip into a depression."[26]

Being "Different" Hurts

But girls are not the only adolescents who suffer from harassment, rejection, and depression. Gay teens of either gender face all the embarrassment and confusion that other adolescents must deal with. In addition, they have to deal with prejudice, stereotyping, verbal insults, and physical attacks. The pain of feeling different, and of fearing rejection from their parents, takes a heavy toll. Government studies have shown that teens struggling with their sexual identity are three times more likely than other teens to commit suicide.

Adolescents with learning or behavioral disorders also have to deal with the pain of being different at a time when most teens crave acceptance. Those with learning problems—dyslexia or attention deficit disorder are examples—get depressed because they cannot do what their classmates can. Besides facing rejection from their peers, they frequently have to deal with parental rejection. Many psychologists believe that learning disorders are greatly underestimated as sources of teenage depression.

Loss and Pain

Significant loss is another major trigger for teen depression. Julia Thorne, now a poet and writer, was literally crippled by her father's death. A few months after he died she was unable to walk

A family enjoys a meal during a break from moving their things into a new home. Relocating to a new town can be very stressful for teens.

without crutches—even though doctors could find nothing wrong with her. "Slowly," she recalls, "my therapist helped me see that my physical pain and handicap was the only language I had for the emotional pain of my father's loss."[27] Julia's reaction was an extreme one, but the death of a parent, sibling, or even a pet can plunge a vulnerable adolescent into depression. For example, when John's twelve-year-old brother died of cancer, he did not cry or show signs of sadness. But in the weeks afterward he withdrew from his friends and family and kept getting into fights at school. It took several sessions with a social worker before John realized that his change in behavior was "because I'm walking around with a big, hurting lump inside me. I tried to pound it out by fighting, but it didn't work."[28]

Losses that lead to depression do not have to involve a death. A breakup with a boyfriend or girlfriend can be just as devastating. Moving to a different school or another town can be extremely difficult for any young person because it often means the loss of all one's friends. Making new friends can be extremely daunting for all but the most extroverted teens.

Two other kinds of losses that are rarely considered as triggers for depression are the loss of childhood and the loss of normality. When puberty begins, it means the end of childhood—whether the teen wants it to end or not. Also, serious illnesses or injuries can separate teens from their peers and make them feel different.

Seventeen-year-old Christopher developed epilepsy after he hurt his head in a skateboard accident. He says it was the absolute worst thing to ever happen to him:

> There isn't an hour in any day that I don't think about epilepsy. I think about it when I see my friends all getting their driver's licenses and know that might not happen for me. I think about it when I'm at the beach with my friends and we're out in the water. Oh, I mustn't go too far . . . might have a seizure. It has made it tough for me to have friends. Some have been great, but some think I'm weird because of the epilepsy, and they're like afraid to know me. I feel that epilepsy keeps me from being normal and being young.[29]

College Blues

Strangely enough, healthy teens who achieve all their goals are often victims of depression, too. The senior year in high school can be a difficult time for kids who have been accepted into the college of their choice. "For years they've talked about and planned for college," says Lee Robbins Gardner, a psychiatrist who deals with teens.

College freshmen away from home for the first time often have difficulty adjusting to life on their own.

"Now they're in and [they] feel frightened and conflicted when confronted with the fact that their dreams are becoming reality."[30]

In fact, going away to college is one of the biggest triggers of teen depression. It involves separating from one's parents, figuring out a direction in life, and forming relationships with a whole new group of peers—both intimate and social. And all these significant challenges take place far away from the support system of home and friends. It is perhaps not surprising that a 2000 survey by the American College Health Association found that a majority of college students admit to feeling "intense hopelessness"[31] at least occasionally.

While the stress of college undoubtedly contributes to depression, the reverse may also be true. Like all depressed adolescents, college students who are truly depressed tend to make bad decisions that only make their situation worse. They may skip classes, avoid friends and family, abuse drugs or alcohol, neglect their diet or engage in risky sexual practices. All these behaviors can increase the stress they feel and contribute to the downward spiral that is typical of severe depression. By the time depressed teens seek professional help, they frequently have a host of other problems. Very often they will be victims of eating disorders, anxiety disorders, personality disturbances, or substance abuse.

Chapter 3

Helpless and Hopeless: The Signs of Depression

ALTHOUGH THERE ARE underlying similarities in the despair and hopelessness—and even the brain chemistry—of teen depressives, the symptoms they display to the world can vary widely. In fact, teen depression is difficult to recognize and treat because it wears so many different disguises. To cite just one example, depressed adults are likely to have trouble sleeping and will often lose their appetites. Depressed teens, on the other hand, often behave exactly the opposite. They are likely to eat and sleep *more*. Yet, since adolescence is a time when growing bodies need more food and sleep, it is hard to differentiate between what may well be a normal part of the growing process and the onset of depression.

Regardless of what depression looks like from the outside, it is associated with subtle but important changes in the way adolescents' brains work. These alterations can lead to more obvious mental, emotional, and physical changes—which is why depression is rightly considered a disorder that affects the entire body.

Trouble at the Gap

The human brain is a remarkably complex organ. Every second it receives and transmits billions of messages along the body's nerve pathways—messages that monitor and control all of the body's senses and functions. Between each pair of nerve cells, or neurons, there is a tiny gap called a synapse. Special chemicals called

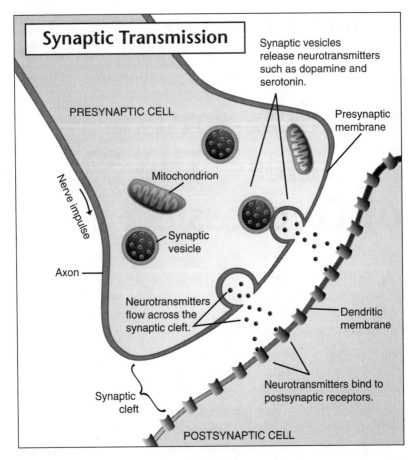

neurotransmitters are critical in moving information across the synapse. There are many different kinds of neurotransmitters, but three of the main ones are norepinephrine, dopamine, and serotonin. Their job is to "bridge" the gap between neurons so that information can efficiently travel to and from the brain.

But if, for example, there is not enough serotonin available for release into the synaptic gap, millions of messages will not be properly transmitted. Such disruptions are strongly associated with depression. Some experts believe that chemical imbalances in the brain actually cause depression. Others are not so sure. They believe that the imbalances are the result of other influences such as the environment, heredity, stress, or the way a person reacts to stress. It seems most likely that depression is caused by a combination of factors.

Whatever the case, when the brain does not work efficiently, the consequences are serious. Since the body depends on the brain to transmit signals of hunger or sleepiness or how to move through the world, all kinds of problems result when communications are impaired. Normal everyday activities like eating, sleeping, thinking, and remembering can seem impossibly difficult for a depressed teen. Even worse, depression has a negative effect on the brain's limbic system, the part of the brain that regulates a person's emotions and motivation.

Apathy, Anxiety, and Anger

"Nothing seemed to matter" is a typical recollection of adolescents who have recovered from a bout of depression. Paul, a Chinese American, remembers his depression as a time of emptiness: "Depression for me was a whole lot of apathy. There were days when I didn't care about anything. The worst part was—well, I'm more intellectual than emotional, and not being able to do anything with my mind was really bad. I felt I was not growing at all. I was stagnating, and it was really unbearable. It made me want to scream."[32]

Besides apathy, depressed teens are likely to be scared and anxious about life—whether they are aware of it or not. For David, that fear was more obvious than for most. In seventh and eighth grades, he always sat in the back of the classroom, shaking and crying with fear. Teased since third grade for being overweight, David hated school and stayed away as often as possible. "I couldn't tell anyone what I was afraid of because I didn't know," David recalls. "No matter how hard I tried, I couldn't control the fear."[33] Starting high school frightened David even more because he knew he would have to interact with older kids. Unable to face that prospect, he dropped out of school. But then David went to see a psychiatrist, who told him that he seemed to have major depression and a chemical imbalance in his brain. With medication David's fears eased, and he enrolled in an alternative school where he fit in much better. "Today," he says, "I'm the same nice, large guy that I've always been—minus the tears, anxiety, and depression."[34]

Other teens express their depression in ways more volatile than David's, and as this mother's recollection indicates, it can be extremely difficult for parents to realize the source of the behavior: "Billy started to become really agitated and restless. He would pick on his younger sister, scream at everyone, and even throw things. I had no idea he was depressed. I just thought he was turning out to be a troublemaker."[35]

"When I get mad, I feel like I can tear down buildings and shred people to bits," Henry confessed. "I can't control my behaviors."[36] As a boy, Henry had witnessed his parents' double suicide. Afterward, he turned the anger he felt toward them (and toward himself) to the world around him. Although bright and articulate, Henry fought continually with his teachers and his foster parents. He blamed others for his misfortunes and refused to follow rules and regulations. Not until Henry was placed in a residential treatment center did he begin to realize that he had never worked through the grief and anger he felt for being abandoned by his parents.

Handling Hurt in Different Ways

Betsy's anger was of great concern to her father and stepmother. She punched her sisters often, got in constant arguments with her family, and even ripped the phone off the wall one night when her

Some depressed teens express anger through violent confrontation, lashing out and blaming their unhappiness on those around them.

father asked her to end a call. After her parents insisted she enter treatment, it became clear that Betsy had not been sleeping well for months. Her underlying problem was that she was depressed and dealing with issues of abandonment. Her real mother, an alcoholic, had left her when she was small and now she feared that her longtime boyfriend was about to break up with her.

A Downward Spiral

Deep inside, teens like Betsy and Henry may feel shame and guilt about their behavior. Although they seem angry at the world, much of that anger is directed at themselves. But those negative emotions only worsen their depressions. Understandably, depressed teens are very pessimistic people. They sincerely believe that things will never get better for them. In fact, unrelieved pessimism is one of the major warning signals for depression, says Harold S. Koplewicz, director of the New York University Child Study Center. "Adolescents who are very pessimistic about how others view them and the way they view themselves—they think they're ugly or stupid and that people don't like them—are more at risk."[37]

Unfortunately for those who would be willing to help, depressed teens tend to hold their feelings inside—particularly feelings of anger, sadness, and disappointment. But doing nothing with strong feelings can make a person believe they have no control over their lives or what happens to them. And people who feel powerless are at a much higher risk for depression.

Suffering in Separate Worlds

Another major reason that the signs of depression can be difficult for parents, teachers, and friends to see is that teens live their lives in three different worlds: home, school, and with friends. Kids who are depressed will function poorly in at least one of these domains but may seem fine elsewhere. Parents are often unaware of problems at school or with friends, while teachers and friends may never have a clue that a teen is dealing with a stressful situation at home.

It is relatively easy for adolescents to keep the beginning stages of depression a secret. Although Joe was seventeen when his mother died in a car accident, he seemed to adjust well. He kept his grades up, continued to see his many friends, and was conscientious about taking care of his younger brother. Even when he had to move to a different school because his father could not stand living in the same house, Joe seemed to be handling a difficult situation well. No one noticed that he seemed exceedingly driven to do well in sports. Besides spending long hours running alone at the track, Joe signed up for an independent study course that kept him up late at night doing research on the Internet.

Yet, right after he was accepted for college Joe drank half a bottle of vodka and climbed into his car with the intention of driving head-on into the nearest telephone pole. Fortunately, his father caught him weaving out of the driveway before Joe could kill himself. Because he was such a "good kid" Joe succeeded in concealing how upset he was about his mother's death and the other upheavals in his life. After being treated for depression, he recovered.

The Signs of Depression

Despite Joe's success in hiding his true feelings, there are a number of reliable signs that an adolescent may be depressed. Perhaps the most important is the length of the symptoms. Adolescence is a time when the chemicals known as hormones are causing major changes in the brain and bloodstream. Therefore it is completely normal for teens to go through periods when they display mood swings or feel negative and stressed out. These feelings are normal and usually temporary. Depression, however, does not go away.

If a teenage girl's boyfriend breaks up with her, she may be crushed. Feeling sad or even heartbroken in such a situation is normal. But if the sadness persists longer than two weeks—if she loses her sense of humor, withdraws from her friends, has disturbed eating or sleeping habits, or cannot keep up with her schoolwork—the breakup may have triggered an underlying depression.

Frequent sadness, tearfulness, or crying are obvious signs of depression, but an inability to enjoy activites that once were pleasurable is also a tip-off. Because Lori seemed "down" her parents decided to raise her spirits by getting her a present of her own choosing. Lori requested a puppy. When it arrived, Lori's joy was short-lived. "Even when we first got the puppy," recalls Lori's

As a result of their low self-esteem, depressed teens tend to isolate themselves from their peers and sink deeper into depression.

mother, "Lori was happy for five minutes and then went off to her room and closed the door."[38]

Depressed teens seem bored with everything. Like Lori they have low energy and tend to withdraw from family and friends. Part of the reason they isolate themselves is that they have such low self-esteem. One deeply depressed thirteen-year-old girl confided to her diary, "Absolutely no one in the world likes me—not even my pets."[39]

Depressed teens often turn to alcohol to help cope with their condition. Unfortunately, alcohol typically exacerbates the symptoms of their depression.

Not surprisingly, low self-esteem often goes hand in hand with an extreme sensitivity to criticism, rejection, or failure. In that fragile state, a poor grade or an insulting remark can seem like too much to bear. That is why some depressed teens erupt with anger or hostility when their behavior is questioned. Adolescents who are depressed will frequently skip school. Often they will complain of stomachaches or headaches. Being afraid to be around other people is called social phobia, and it may be a precursor of depression. "Today we know that social phobia should be seen as a red flag for depression,"[40] says Harold S. Koplewicz, an expert in the field.

But even when depressed teens attend school regularly, they often do poorly because of an inability to concentrate—another common sign of depression. That inability may be further hampered by drug or alcohol abuse, which tends to make the problem worse. Alcohol, after all, is a depressant—it slows down the nervous system and may intensify sadness and lack of energy. Somewhere between one-third and one-half of the teens who abuse alcohol suffer from either mild or major depression. In essence, they have *two* problems instead of one. And, drug or alcohol abuse can mask the symptoms of depression, making it harder to identify and treat.

Sleep disturbances are the number one complaint of depressed people—depressed teens may sleep too much or too little. The tip-off is a disruption in normal sleeping patterns. The parents of one fourteen-year-old, who was later diagnosed with depression, discovered that he was staying up till three in the morning playing computer games. He still did well in school, but when he got home in the afternoon he was so tired he would go right to bed and sleep through the dinner hour. Then he would eat late, start his homework late, and go to bed late again.

Their Own Worst Enemies

Self-destructive behavior is another common characteristic of severely depressed teens. Sometimes it takes the form of accident proneness or neglect of one's personal appearance. For others, self-mutilation is a way of coping with unbearable emotional pain. Raquel, an attractive seventeen-year-old with wealthy parents, appears to be a well-adjusted teen at first glance. But when she rolls up the sleeves of her sweater, she reveals a series of ugly scars carved into her arms with a jagged piece of glass. Raquel has never felt accepted within her family and has used drugs and alcohol since the age of nine. Inside she feels sad and wounded. "Ever since I can remember," she says, "my heart has hurt."[41]

Julie, a depressed teen, explains why she mutilates herself:

> I'm trying to hide the carefully scabbed slash marks on my ankle from my mom and dad for fear they'll hide all the kitchen knives like they did the last time I cut myself. I think that the cutting has been the hardest part of my depression for my parents for the simple reason that they don't understand.... They just don't get it. It's so hard to rationalize pain. Cutting lets off pressure. Some days I must cut or I will explode, swelling up like a manic balloon.[42]

Suicide, of course, is the ultimate self-destructive behavior and talk of suicide is an important warning sign of depression. It should always be taken seriously. Although thinking about death and suicide is not uncommon for teenagers, statements like "I'm going to kill myself" should never be ignored. Still, even such straightforward statements do not always mean that the speaker is depressed.

A Disease with Many Faces

One of the great difficulties in diagnosing depression is that no single sign, by itself, is a sure indication of the disease. The clustering of symptoms and their persistence over time is the key to identification. Quite often it takes an expert on teen depression to see it. Where depressed adults tend to feel sad and empty, teens are more likely to be irritable, hyperkinetic, and aggressive.

Another major difference between adult and teen depression is the instability of teen depression. "While a 40-year-old is going to stay glum until the depression lifts, you can't count on a teenager to stay depressed 24/7," says Koplewicz. "She has an ability to snap out of it, even if it's just for a few hours when she goes out with her friends, before falling back into depression."[43]

Symptoms of teen depression can also mimic normal developmental behaviors. Anger and withdrawal are often part of becoming independent from one's parents. Therefore, behaviors like arguing with parents, refusing to do chores, trouble with teachers, sleeping a lot, and changes in weight may or may not be indications of depression.

Is It Really Depression?

Further complicating matters, many physical conditions produce symptoms similar to depression. Alcohol or drug abuse, hormonal imbalances, as well as reactions to prescription drugs can all produce symptoms of depression. These need to be ruled out before treatment can begin. In addition, complaints frequently associated with depression—like the headaches and stomachaches that keep some depressives from attending school—can fool doctors into thinking the problem is only physical. Because there is a shortage of doctors trained to identify problems specific to teens, depression is frequently misdiagnosed.

Helpless and Hopeless: The Signs of Depression

The stigma attached to mental disorders also makes parents, teens, and doctors more likely to attribute a problem to a purely physical cause. Many parents are reluctant to even consider a diagnosis of depression. In their eyes that would mean admitting that they are bad parents—even though that might not be the case.

Because depression begins silently and slowly, teens' parents and friends are often unaware of its onset. Only when it becomes severe do others take notice. Teens who are depressed often do not realize it themselves. They are hit emotionally with so much during adolescence that some of them develop an intolerance for their deepest feelings. Boys, especially, are unable to articulate their pain. Instead, their emotions are discharged through their actions. And, finally, even teens who realize that they have a problem are reluctant to admit it to their parents because asking for help can feel like a regression to childhood.

Hurdles like these can make it difficult to identify teen depression. Even when it is properly diagnosed, there are at least four

A teenage girl has an emotional talk with her mother. Many teens suffering from depression are unable to articulate their feelings.

distinct varieties of the problem. The mildest and most common is reactive depression. It is called "reactive" because it is characterized by difficulty in adjusting to a disturbing event—something as serious as a death in the family all the way down to a minor rejection. A reactive depression lasts from a few hours to a few weeks and is not considered a serious enough problem to be a true mental disorder.

A World Without Joy

Dysthymic disorder, on the other hand, lasts an average of four years in adolescents. Also called neurotic depression, it could be characterized as living life in slow motion with a persistent lack of joy. Teens with dysthymia can get through their daily routines, but it takes them great effort to do so. Poor appetite, low energy, insomnia, the inability to concentrate, and feelings of hopelessness are typical with dysthymia.

It is a condition that can have serious repercussions for a young person's future. Since dysthymia interferes with the ability to experience pleasure, it impairs the kind of feedback crucial for normal development—teens who do not know what makes them happy do not really know who they are. Getting help early is important since experts believe that up to 70 percent of young people with dysthymia will go on to develop major depression.

Severe Depression

Serious or severe depression is also known as major depressive disorder (MDD). Adolescents with MDD have great difficulty with everyday functioning. On average, bouts of severe depression last from seven to nine months. They are characterized by a significant change in mood lasting for more than a few weeks. That change is accompanied by the presence of a number of other symptoms.

These symptoms last most of the day, nearly every day, and sometimes include appetite changes or sleep disturbances as well as a loss of pleasure in most activities. A severely depressed teen is likely to feel sad, empty, worthless, anxious, or hopeless. Sometimes the negative feelings will take the form of extreme guilt.

Parodoxically, depressed teens may display either extreme apathy or extreme agitation. Often they are unable to concentrate

Adolescents suffering from severe depression, or major depressive disorder, have difficulty functioning in day-to-day life.

and are extremely fatigued. Loss of energy to go about one's daily life is a strong sign of depression. So are recurrent thoughts of death or suicide.

In order to be diagnosed with MDD a teen would have to display at least five of the above symptoms. The diagnosis can be tricky because teens with MDD are more anxious and irritable than adults with the same problem. Frequently, they have an atypical version of MDD that is characterized by being overly sensitive to rejection and failure. Often they will act in ways not normally thought of as depressive—being aggressive or sleeping and eating too much, for example.

Elation and Deflation

Researchers have reported that between 20 and 40 percent of teens with MDD will go on to develop bipolar disorder. Sometimes called manic depression, bipolar disorder is characterized by alternating swings in mood and energy. Like MDD, bipolar disorder does not always look like what most people think of as depression, therefore it can be tough to diagnose. When a teen with bipolar disorder is in the manic stage, he or she may be elated, excitable, creative, and confident. "Sleep was impossible for me," recalls Amber. "I'd dance, sing and write poetry all night, then go to school in the morning."[44]

Manic depressives who are "up" can be great fun to be around. But when they are "down," their symptoms are usually indistinguishable from major depression. Bipolar disorder recurs often—90 percent of those who have a manic episode will have another—and accurate diagnosis is critical because the medications involved in treatment are different than those for severe depression. About a fourth of manic depressives will one day attempt suicide. Like all the depressive disorders, bipolar disorder can do serious and even permanent damage to a developing teen.

Chapter 4

When Depression Turns Deadly

SUICIDE IS THE most extreme manifestation of severe depression. Each year it takes the lives of more teens than all other illnesses combined. But, as with depression, a potentially suicidal person can be treated successfully if correctly diagnosed. Far too often, though, suicidal teens do not get the help that can save their lives.

A week before Christmas in 2001, a slim, attractive fourteen-year-old named Marissa Imrie left her second-period class at Santa Rosa High School near San Francisco. A straight-A student, Marissa took a taxi ride to the Golden Gate Bridge. Once there, she climbed over a barrier and jumped to her death in the swirling waters 220 feet below.

Although Marissa had recently complained of headaches and insomnia, her mother, Renee, with whom she was exceptionally close, had no idea her daughter was depressed. On Christmas Day Renee was going through Marissa's things when she found her suicide note. Its despair and distorted reasoning are typical of severe depressives. It ended with this plea: "Please forgive me. Don't shut yourselves off from the world. Everyone is better off without this fat, disgusting, boring girl."[45]

Psychologist Calvin Frederick, an expert on depression, says teens who seriously consider killing themselves tend to have three things in common: "When a teenager is suicidal, the three H's—helplessness, haplessness and hopelessness—are usually present. Events seem to be conspiring against the child, who feels helpless to deal with the problem. There seems to be no hope that

A severely depressed teen girl sits on the floor with a knife beside her as she contemplates committing suicide.

things can be different. It is this loss of hope which can lead to serious depression and, possibly, suicide."[46]

Jason, a seventeen-year-old who works at an Arizona crisis center, reminds suicidal people that taking their life is a "permanent solution to a temporary problem. I always ask," he says, "Who would miss you if you weren't around?"[47]

Such questions are relevant because severely depressed teens find it very difficult to realize how their actions will affect others—or that the present situation will ever change. Again, because of their youth, they usually have had little experience in recovering from painful emotional experiences. Like Marissa they literally cannot imagine things ever getting better. And thinking positively about the future seems to require more mental energy than they possess. A person who sincerely believes that an intolerable situation is never going to end is a person prone to thoughts of suicide. From all indications, that was how Kurt Cobain felt right before he killed himself at age twenty-seven.

Insurmountable Pain

Beverly Cobain, a psychiatric nurse, is Kurt Cobain's cousin. Before he died she tried unsuccessfully to reach him so that he could get help. Since then she has become dedicated to helping prevent other young people from taking their lives. She reminds people that "When a depressed teen says 'I want to die,' all of us need to hear 'I'm feeling so alone (helpless, hopeless, worthless) right now, that it's unbearable. I can't think of any way to escape those painful feelings except to die.' "[48]

The hopelessness and despair Kurt Cobain experienced is shared by millions of severely depressed adolescents every day. Yet there is little understanding of the true nature of suicidal depression. When Andrew Slaby, a psychiatrist specializing in depression, asked people how they pictured a depressed person, they tended to describe a sad, miserable person who shuffles through life before taking an overdose:

> There is no understanding or recognition of the rage, fear, and the insurmountable pain that are so much a part of depression. Imagine the worst pain you've ever had—then multiply it tenfold and take away the cause; then you can possibly approximate the pain of depression. The mental pain of depression is so all-consuming

Volunteers at a teen suicide crisis hotline offer a message of hope to suicidal callers who have reached out for help.

that it becomes impossible to derive any pleasure or satisfaction from life.... The world is seen as bleak and gray. To someone who is profoundly depressed, the option of suicide becomes the only option, the only way to control life and end the pain.[49]

A teen who is depressed is thirty times as likely to commit suicide as someone who is not. If, like Kurt Cobain, the individual is impulsive by nature, the danger is heightened. "Typically," says psychiatrist Harold S. Koplewicz of teens who respond to stress with agitation, anger, or hopelessness, "they think that they are the cause of the negative event; that they have no control over what happens to them; that nothing they can do will change things; and that they will never get better."[50]

Millions in Misery

A recent survey by the Centers for Disease Control found that the number of young people who have thoughts like these is distressingly high. According to the survey almost one out of every five teenagers—about 3 million per year—admitted thinking about suicide. Of those 3 million, 2 million made actual plans to carry out their deaths and some four hundred thousand actually made suicide attempts serious enough to require medical attention.

Of course, contemplating suicide does not mean a teenager will successfully carry it out. Still, in 2002 for example, about two thousand teens did succeed in killing themselves. Suicide attempts become more common around age fifteen and peak around age sixteen. It is thought that the lower rates among older adolescents are because they are a little more likely to realize how devastating their death would be to family and friends.

Although girls are twice as likely as boys to attempt suicide, they are less likely to die. Only one in three thousand girls who attempt to kill themselves will succeed. Boys are ten times as likely to die of a suicide attempt—one out of every three hundred results in death. The gender discrepancy seems to be mostly due to the methods chosen. Girls tend to overdose on drugs, while boys tend to use guns, which are much more lethal. Fortunately, teens rarely attempt suicide without giving some clues as to how they are feeling.

Suicidal teens often signal their intentions to friends before making an attempt at suicide. Here, a depressed teen discusses his condition with a friend.

Warning Signs

Verbal hints are one of the warning signs of suicide. It is quite common for suicidal teens to signal their intentions beforehand in some fashion. A suicidal teen may complain of being a bad person or feeling "rotten inside." He or she may consider themselves a burden to others. He or she will say things like "I won't be a problem much longer" or "I'd be better off dead." Too often, as this journal entry by a teen named Rachel makes clear, teens want to communicate their despair to others but do not quite know how to do it:

> If I'm not doing what I want to do in a few years, then I will commit suicide. I'm surprised I haven't done it yet because I feel really terrible all the time—well, maybe most of the time. I just don't have the courage but anyone else in my position would have already committed suicide.
>
> ... I know when people hear this they're going to say that killing yourself is stupid and I feel like they are yelling at me. I think it makes me feel worse—not that anyone will ever hear this because I've written these a million times to tell people how I feel.

I am hoping that I would show this to someone today finally to tell how I feel. I'm really confused about everything. Usually my friends can help me but now there is nothing they can do.[51]

Other actions that can signal an impending suicide attempt include putting all of one's affairs in order, becoming angry when praised, giving away favorite possessions, or becoming inexplicably cheerful after a period of depression. A preoccupation with death—listening to songs about death or drawing or writing about death—can also be a tip-off. A fascination with death was one of the warning signs for Ryland, a teenager in Washington State who eventually jumped off a bridge (he survived but is paralyzed for life). His mother recalls how shocked she was to learn what was really going on inside her son's head: "One morning, I secretly reached under the futon and drew out a notebook filled with dark, sad, morbid poetry about life not being worth living. I realized Ryland was suicidal, and I couldn't understand why. He was smart and good looking—he had everything going for him."[52]

Suicidal teens often engage in reckless behavior. This teen is buying illegal narcotics on the street.

Suicidal teens tend to take unnecessary risks. They do too many drugs, drive too fast, or drink too much. Alcohol abuse, in fact, is one of the highest risk factors for suicide. Some evidence suggests that the huge rate of increase of teen suicides between the 1950s and the 1980s can be attributed mostly to the rise in drinking rates of young people. That seems plausible since suicide is very often associated with impulsive behavior and there is no question that alcohol reduces inhibitions and increases impulsiveness. Sometimes suicidal teens show signs of psychosis—they will have hallucinations or express bizarre thoughts. In such cases, alcohol abuse will only make the job of dealing with their problems more difficult.

How to Help

Virtually all teens who seriously consider killing themselves have one thing in common: They believe that they are facing problems too overwhelming to solve. The key to helping them is changing that perception. First, however, someone needs to notice symptoms like those mentioned above—and then take action to get help. In some instances that will be a teacher or a friend, but in most cases it will be a parent. According to psychiatrist Harold S. Koplewicz, "the biggest difference between a teenager contemplating suicide and one who actually attempts it is recognition of problems by his or her parents before it is too late."[53]

Suicidal teens are so focused on their own emotional pain that they are incapable of thinking anything good about themselves or their lives. Dying seems like the only way they can escape the pain. Still, deep down, most suicidal teens do not want to die—but they do need help in learning how to cope with problems that seem insurmountable.

That is why just hearing a concerned person say "I want to help" can sometimes be enough to reverse suicidal thoughts. And, it is why asking about suicidal feelings is so important. Maureen Empfield, director of psychiatry at Northern Westchester Medical Hospital in Mt. Kisco, New York, has this to say on the subject: "It is especially important to know if the teenager has thought of harming himself. Parents may be completely unware that their teenager has thought of committing suicide—many kids, even kids who are

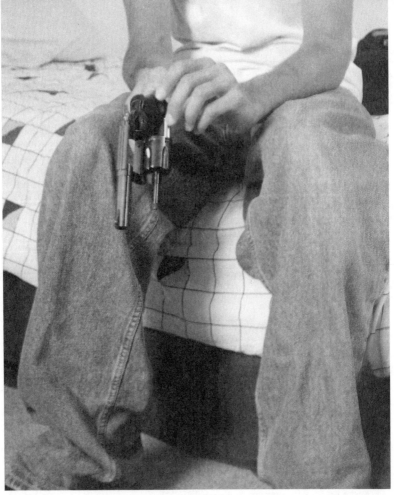

Teen boys who attempt suicide with a gun have a greater chance of actually killing themselves than suicidal teen girls, who typically resort to drugs.

highly verbal and expressive, just keep it to themselves. Asking directly is the best way to find out."[54]

Discussing the situation with someone who cares can help a troubled teen see that there are other options—the problem is almost never as big or as insoluble as first imagined. As James Baldwin, the famous writer, once said, "Not everything that is faced can be changed. But nothing can be changed until it is faced."[55]

If someone is hinting at suicide or threatening to commit suicide, there are a number of things a person can do to help. First, it is important to remain calm and show genuine concern. Tell the potential suicide that you want to help and take his or her feelings seriously. Reassure him or her that you know how to get help, then stay close to the person while you notify a trusted, responsible adult that help is needed. If it is safe to do so, remove items that the person could use to attempt suicide and then remain with the person until professional help arrives.

A Safe Place to Get Well

If a teen is acutely suicidal or has actually attempted suicide, he or she will most likely be taken to the emergency room of a hospital, and, if there are any medical complications, will be hospitalized in an intensive care unit for at least twenty-four hours. But even if there are no apparent physical injuries a doctor will do a medical workup. That is because there are many medical conditions, including acute intoxication, that can cause depression and suicidal impulses.

Once a medical evaluation and any treatment has been administered, it is the turn of the psychiatrist. He or she will be called when the patient is medically stable or is able to communicate. The psychiatrist will talk with the teen, evaluate the situation, and recommend a course of treatment that may include drugs and/or

A physician discusses treatment options with a teenage girl. In addition to medical attention, suicidal teens often receive psychiatric treatment.

therapy. In some cases, psychiatric hospitalization will be recommended. That is what happened to Cait Irwin: "The reason that I needed to check into the hospital was that I got to a point when I couldn't trust myself. I didn't know if I wanted to live or die. I went to the hospital to talk about my problems and to find the right medicine to help me. But most importantly to keep me safe."[56]

A psychiatric ward provides a safe environment for patients who might otherwise try to escape or harm themselves or others. In most psychiatric treatment, both medication and therapy are used together because they seem to work better when used in tandem. Once the right course of treatment has been found, most suicidal teens feel much better. They discover that talking about their problems is preferable to keeping everything locked inside.

The teens most likely to be hospitalized are those who are suicidal. But sooner or later their condition will improve and they will be released. Their treatment, in almost all cases, will continue. They will be asked to continue taking antidepressants and to meet with a therapist or counselor regularly. Often teens will not want to keep their appointments. Since they feel so much better, they will try to convince their parents that the danger has passed. Parents who give in to these pleas may live to regret it.

"This course of action—and there is no gentler way to state it—can be fatal,"[57] says psychiatrist Maureen Empfield. When depression recurs in youngsters who thought they were over it for good, the resulting despair can be life-threatening. Over 40 percent of teens who have recovered from a suicide attempt will try it again within two years. Since the single best predictor of whether a teen will attempt suicide is a prior attempt, it is absolutely essential that teens released from hospitalization continue with any treatments prescribed for them.

Potential suicide, like depression, is a treatable mental disorder. The greatest hope for its prevention is the early diagnosis and treatment of depression.

Chapter 5

Restoring Hope: Treating Depression

ALTHOUGH DEPRESSION CAN be devastating, it is a manageable illness. Both drugs and talk therapy have proven effective in helping depressed teens return to normal functioning. The good news is that teens in general respond well to treatment and have high rates of recovery.

Sixteen-year-old Mackenzie had been a good student, but depression caused her to become discouraged about her future. She began cutting classes and abusing marijuana. She hid the stress she was feeling because she was ashamed to tell her parents and did not want to be a burden to her friends. The key to her recovery was a couple of caring adults who were willing to listen:

> One day, the coach approached me and said that someone on the team had told him I'd been high at one of our games. The dam broke wide open. I was desperate for someone to know what was going on inside me, so I told him everything, and it felt great to talk to someone who cared.
>
> . . . I started to see a therapist and she's been very easy to talk to. Terry doesn't judge what I say, do, or feel. She's objective and listens without getting emotional, and I can take suggestions from her more easily than I can from my parents. . . . She also helps me work through my problems. I'm glad I have a therapist who listens and understands.[58]

Before fifteen-year-old Hilarie began talking with her school's guidance counselor and received treatment for depression, she was tortured by crying spells, low self-esteem, and insomnia. With medication she now feels "normal again." "Take a close look at yourself," says Hilarie. "The way to recovery is recognizing you need help immediately."[59]

The outlook for teens like Hilarie and Mackenzie is good, but for those who do not get help the situation is not as bright. Early diagnosis of depression is essential because those who are not treated are at higher risk for further episodes—episodes that can be increasingly severe. Unfortunately, the majority of depressed teens do not receive the help they need and many will have recurrences as adults. One study found that 31 percent of those who had depression as adolescents went on to have depression as adults (as compared to only 8 percent of teens who had never been depressed).

Talk therapy helped bring Mackenzie out of her depression. Both psychotherapy and counseling are kinds of talk therapy. The difference is essentially in what kind of person does the listening. Psychotherapists are usually psychologists or psychiatrists. Counselors can be any kind of mental health professionals—like social workers, for example. Talk therapies are most effective when the depression is mild or moderate, has not lasted a long time, and does not involve a mental disorder so serious that it interferes with an adolescent's ability to deal with reality.

The Importance of Acceptance

Regardless of what kind of talk therapy is used, one of its most important functions is to help depressed teens realize what they are up against. As psychiatrist Maureen Empfield points out, "Most youngsters (and most adults, too) are reluctant to accept the idea that they have a mental illness of any sort."[60] No one wants to be thought of as "crazy"—even though that is not at all what a diagnosis of depression indicates. In essence, depression is a physical disorder that can be triggered or worsened by psychological stresses—therapy is for people who are feeling emotional pains and need an objective listener.

Talk therapy is most effective with teens suffering from moderate depression. Here, a teen girl has a session with a counselor.

Once the diagnosis of depression is acknowledged and its effects recognized, teens are much more likely to take any medicines that are prescribed or show up for therapy with a positive attitude. The best therapists may even get their patients to become curious about what the process of their therapy might uncover.

Twisted Thinking

Among the many different kinds of therapies used with teens, one of the most useful is cognitive therapy. Cognition is a word that refers to how a person sees the world. The idea behind cognitive therapy is that depressed people tend to see the world in a distorted way. Their thinking tends to be overly negative or illogical in ways that are apparent to others, but not to themselves. If such thinking persists long enough it can lead to the kinds of physical impairments associated with depression.

A boy who does not get into a prestigious college, for example, concludes that he is a complete failure and always will be. A girl whose boyfriend breaks up with her decides that she is ugly and fat and will never date again. After doing poorly on a test, another teen concludes that she is incapable of ever learning. Sometimes depressed teens walk around in a gray cloud of depression because of unnecessary guilt. Here's how a ninth-grade teacher describes one of his students: "Nathaniel walks around as if the weight of the world is on his shoulders—he feels personally responsible not only for his parents' breakup, but for every argument or quarrel in this classroom."[61]

Sometimes the distorted thought can be as straightforward (and as debilitating) as "I feel bad, therefore I must be a bad person." Cognitive therapists believe that this kind of thought only leads one deeper into depression. "If it's true that thinking determines mood," says psychiatrist Harold S. Koplewicz, "depression may be treated or prevented by intervening in a child's distorted thinking patterns before they become embedded in the brain."[62]

Cognitive therapy generally works better in milder cases of depression, and in severe cases—the kind that also require medication—it seems to be helpful in preventing relapses. One of the pluses with this kind of therapy is that it is relatively short-

A teen girl appears depressed after arguing with her boyfriend. Some teens become severely depressed after ending relationships.

term. The ultimate goal is to get teens to feel better about themselves and, of all the therapies, it seems to work fastest at combatting depression. However, if improvement is not seen within ten or twenty sessions, it may be time to try something else.

Other Useful Approaches

Interpersonal therapy, or IPT, is a problem-oriented kind of talk therapy often used with teens. Its goal is to improve the depressed person's relationships with the important people in his or her life. This involves improving communication skills and concentrating on specific issues that the teen is currently facing—a conflict with a boyfriend, an abusive parent, or problems at school. IPT therapists try to relieve depressive symptoms by solving interpersonal problems. Once-a-week sessions last for three to four months, but can go longer.

Good counselors and therapists will use whatever works best with a particular teen. That might be cognitive therapy and IPT—especially if changes in behavior are deemed helpful. But, if unconscious motivations are suspected, then more typical psychoanalytic techniques can also be used. These might include talking about the past or one's dreams. Although many people think of all therapy as a one-on-one encounter between patient and therapist, other methods involve more than just two people.

A family therapy session might include an adolescent's parents and siblings. Depressed teens need support and understanding and there are times when the best way to ensure that they get it is to involve the whole family in the process. When John was sixteen his family moved to a foreign country for a few years. Because John was attending a local school he picked up the language much faster than his parents. It was a problem that only became apparent when the three of them went to a family therapist to find out why John was so depressed:

> His parents began to rely on John for almost all the interactions with local agencies, neighbors, and so forth. At first, John felt proud and helpful, but soon he began to demonstrate resentment and then depression for having to be the "voice" for his

Family therapy sessions often involve the entire family to provide depressed teens with the support and understanding they need.

family. When the parents consulted with a family therapist, the dynamic was pointed out, and the parents became more aggressive in learning the language and in staying within their defined roles as parents.[63]

In group therapy depressed teens get together with others like themselves to discuss their situations. Sometimes just realizing that other adolescents are dealing with similar issues can give a young person valuable insight into his or her own problems.

Someone to Talk To

No matter what kind of talk therapy is used, it is crucial that the therapist be a person the teen feels comfortable with. Not all adults relate well to teenagers. If, after three sessions, a bond of trust and understanding has not developed, then it may be necessary to find someone who is a better match. It is important, however, to keep in mind that opening up to a stranger and discussing painful feelings is difficult for almost anyone. Having doubts about the process is natural, particularly in the beginning.

Even when therapy seems to be working well, therapists caution against ending it too soon. Once depressive symptoms have been eliminated, relapses are possible and the consequences of their return can be devastating. "The sense of failure I felt in having it come back was crushing, and no doubt contributed to making this new episode of depression worse,"[64] recalls fifteen-year-old Anna.

Continued therapy helps teens build skills that can prevent future setbacks. Tyler was an angry teen before he was helped by a counselor that he continues to see:

> When I was depressed, I would let myself think about how I was messing up in school and other things, and I'd just crumple up into a ball on my bed. My insides would churn and squirm. I didn't know what to do or how to think about what was happening. I wanted to escape—to press a button and be out of my body. I didn't want to feel. When I got angry, I either exploded or walked away. Counseling finally helped me. . . . The more I talked to the counselor, the more I could organize my thinking. It was good for me.
>
> . . . The new emerging me, the me right now, is very mellow. I don't have blasting anger, and I don't feel depressed.[65]

The process of therapy is not something one need be ashamed of. It is hard, courageous work that has helped millions of people all over the world. Tyler describes the therapeutic relationship most likely to lead to healing this way: "I believe we find our own answers to our problems, but sometimes we just need help asking the right questions."[66]

Mood Medicine

Sometimes, however, talk therapy is not enough. In recent years, a new class of drugs called antidepressants have revolutionized the treatment of depression. "After three months of being severely depressed, I went on Prozac," says fifteen-year-old Alex of Jonesboro, Maine. "I've been on it for two years, and it has helped me get my life back."[67] Antidepressants change the amounts of brain chemicals like serotonin that regulate thinking, emotions, and behavior. Although the way they work is still not completely understood, antidepressants do bring teens out of depression and help prevent future depressions.

Although these drugs may improve emotional control and can have a positive impact on psychological symptoms, their main effect is to get adolescents' brains and nervous systems back to working normally. Antidepressants are not tranquilizers, they do not make people high, and they are nonaddictive. Most teens who respond well to antidepressants for a first episode of depression will only have to take the drugs for six months to a year.

Antidepressants are most often prescribed when the teen is experiencing severe depression, has suffered previous episodes of depression, or comes from a family with a history of depression. They might also be tried when the adolescent does not have access to talk therapy or cannot afford it.

Like almost all drugs used on human beings, antidepressants can cause side effects. Some studies have suggested that, in rare cases, antidepressants increase the risk of suicide. More research will need to be done if before such a connection can be confirmed. Meanwhile, it is important that any drugs used be prescribed by a child psychiatrist or other professional who has experience treating teens. In most cases, however, the side effects will merely be unpleasant. Typical side effects are nervousness, diarhhea, dry mouth, nausea, headaches, and sleep problems. They can usually be minimized or eliminated by reducing the dosage or using a different antidepressant.

Getting It Right

These kinds of adjustments are made during the first, or acute phase, of depression drug treatment. This phase is a process of trial and error that requires cooperation between the patient and doctor until the right drug and dosage is found. Communication is important because antidepressants usually take four to six weeks to start working. "It's like the neurotransmitters in your brain have been hibernating like bears and are slow to wake up," says psychiatric nurse Marcia Bentley. "The medication has to build up to a level where it actually relieves your symptoms."[68]

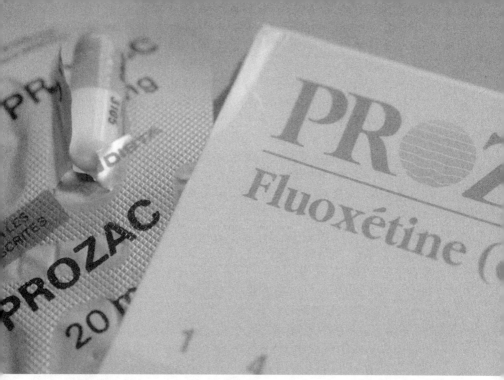

Prozac belongs to a class of drugs known as antidepressants, which regulate the brain chemistry of depressed individuals.

All this may take a while, because if one drug does not work another will have to be tried.

Once an effective medication is found, the maintenance phase of the treatment begins. In this phase the main goal is to make sure that the patient takes the pills. That is important because antidepressants are not effective unless they are taken regularly.

Depression-Fighting Duo

Although antidepressants have become much safer and effective in the last decade, they are usually only part of the treatment. Except in mild cases of depression, therapy and medication seem to work better together than either does alone. While antidepressants can produce amazing results, they are not wonder drugs. They cannot, for example, solve the kinds of life problems depressed teens often face. Therapy is much more effective in dealing with those kinds of issues. On the other hand, without medication to limit some of the physiological symptoms of depression, therapy can be difficult or impossible. Both treatments are valuable—and each usually works better when combined with the other. Treatment that combines medication and some form of talk therapy is particularly beneficial when depressive episodes occur in rapid succession without a full recovery in between.

A Safe Environment

In cases of severe depression—the kind that involves suicidal thoughts or psychosis—a depressed adolescent may be hospitalized before any kind of treatment can begin. What usually happens is that teens are brought to an emergency room by family or friends as a result of frightening behavior: They threatened to harm themselves or others. Such behavior is almost always grounds for hospitalization—whether the teen agrees to it or not.

After any injuries are treated and the patient's physical condition is stabilized, a psychiatrist will discuss the situation with the patient and his or her family and decide on a course of treatment. In most cases the goal will be to stabilize the patient so that further treatment can be given outside of a hospital setting. In some

Although antidepressants can seem to lift teens out of depression, talk therapy is still needed to help depressed teens work through issues.

Restoring Hope: Treating Depression

instances, however, treatment is best given in a setting where the patient's condition can be closely monitored.

Lee, a college freshman, was having great difficulty adjusting to being away from home. Severely depressed, she was unable to go to classes and rarely left her room. Finally, the school sent her home. Her behavior, even though she was seeing a therapist, continued to worsen. She refused to leave the house and, although she had been prescribed antidepressants, no one knew for sure whether she was really taking them. Finally, Lee and her family agreed with her therapist's recommendation that hospitalization was the best course. Once Lee was in a setting where her drug dosage could be monitored, she quickly got better.

Although hospitalization can literally be lifesaving, too many people think of it as something shameful. A teen who goes into a hospital because of a medical disease is likely to receive sympathy from family, friends, and schoolmates. But a teen who is hospitalized for a psychiatric illness may be thought of as crazy—even by his or her own family. Such attitudes can keep teens from getting the treatment they need to return to health.

The Busy Adolescent Brain

Recent research in the area of depression provides new hope that depression treatments will be more effective in the future. Jay Giedd, a researcher at the National Institute of Mental Health, conducted some intriguing studies on what goes on inside the adolescent brain. Previous research had shown that the brain's gray matter—the part of the brain where higher intellectual functions are carried out—gradually thins out until about the age of twenty-five. Giedd discovered that gray matter continues growing in girls until about age eleven and in boys until age twelve. Suddenly, just as adolescence is beginning, the brain begins "pruning" unused brain connections.

This reorganization of the brain seems to coincide with an increased ability to do specialized kinds of thinking. It is as if the brain is ridding itself of unused connections in order to become more efficient. Giedd's most interesting finding was that brain reorganization peaks between the ages of fourteen and seventeen—it

Inside the Adolescent Brain

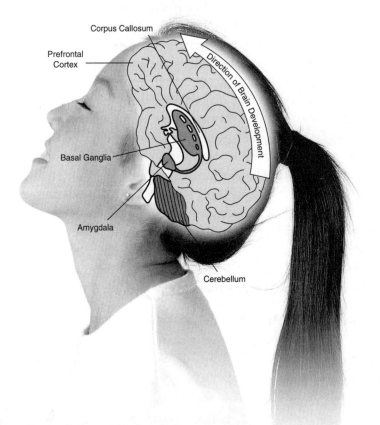

Corpus Callosum: During adolescence, this bundle of nerve fibers thickens and processes information more and more efficiently.

Prefrontal Cortex: Area of practical thought and last part of the brain to mature.

Basal Ganglia: Helps the prefrontal cortex prioritize information. Active in small and large motor movements.

Amygdala: The emotional center of the brain, home to such primal feelings as fear and rage.

Cerebellum: Long thought to play a role in physical coordination. Supports activities of higher learning like mathematics, music, and advanced social skills.

occurs at a rate as much as four times higher than at any other time in life. This means that adolescent brains are undergoing far more restructuring and change than previously thought.

Interestingly, the burst of brain activity between fourteen and seventeen also coincides with the time when adolescent psychiatric disorders like depression are most likely to begin. Another of the country's leading researchers on child psychiatry, Adrian Angold, believes that hormones associated with puberty somehow activate genes that make teens more vulnerable to depression or anxiety. All this suggests that there might be biological reasons why some teens are susceptible to depression—reasons that might one day be understood so that remedies can be devised.

A Test for Depression?

More evidence to support this idea has come from teens who, after attempting suicide, volunteered to undergo spinal taps for research. Those taps revealed that they had unusually low levels of serotonin in their blood. Serotonin has many functions besides regulating one's mood—blood clotting, digestion, and sleep are just some of the others—so the findings do not necessarily mean that low serotonin causes depression. It could be an effect rather than a cause.

But the findings do suggest that researchers may be on the verge of finding biological "markers" for depression. Perhaps one day soon there will be blood tests or other kinds of screening tests that can identify those teens most at risk for getting depression.

Meanwhile, increasing numbers of antidepressants have been discovered. The newer ones have fewer side effects and are safer—even if an overdose is taken they are not likely to be lethal. Talk therapy has become more effective, also. Studies have shown that cognitive therapy combined with relaxation training or group problem-solving can offer long-term protection from depression. That may be one reason that suicide rates for teenagers have decreased in the last ten years. Since the most effective treatments for major depression are also the best defense against suicide, it would appear that the increased use of new techniqes and new treatments is beginning to have a positive effect.

Chapter 6

Keeping the Blues at Bay

SUFFERING LOSS AND dealing with painful situations are part of the human condition. Every adolescent feels sad and down on occasion. But more effective treatments and a new generation of antidepressants make it likely that even those teens most vulnerable to depression can find ways to alleviate their emotional pain. Meanwhile, increased awareness of the true nature of depression can motivate teens and parents to learn new behaviors that will keep depression from returning when the next challenge arrives.

Stress is inevitable—perhaps never more so than during adolescence—yet emotional responses to stress vary from individual to individual. Everyone who lives in New York City was stunned by the terrorist attacks on the World Trade Center on September 11, 2001. But most teens, including those living nearby, handled the aftermath well. Yet, some teens in other parts of the country who may only have witnessed the attack on TV became so incapacitated by fear that they refused to go to school.

The way a teenager manages stress determines whether they are at risk for depression. "There is no single cause, single treatment, or single defining characteristic," advises psychologist Michael D. Yapko, an international expert on depression. "So don't try too hard to look for either THE cause or THE solution."[69] The key to effectively dealing with depression is an awareness of its nature. Yapko emphasizes that the kind of guilt that increases hopelessness and inertia in so many depressed teens is not helpful. "Depression is more accurately considered a disorder than a disease. There are biological influences underlying depression,

Keeping the Blues at Bay

just as there are sociological and psychological ones. You are not a diseased or sick person; you *are* someone who has not been taught to develop the specific ways of thinking, feeling, or behaving that [will] insulate you against life's difficulties."[70]

Parents, too, may need to be "insulated" from the shame associated with depression. Too often they conclude that what has happened to their children is their fault. That is not necessarily true, say clinical psychologists Laura Epstein Rosen and Xavier Amador, "It could happen to anyone. Your child may have a genetic vulnerability that came from your mother's side two generations back and has nothing to do with the kind of parent you have been."[71]

This teen girl stays in touch with her feelings by keeping a journal. This behavior helps her cope with everyday stresses.

Looking Ahead, Not Back

Rather than figuring out cause or blame, it is much more helpful to focus on ways to get depressed teens well and keep them that way. "The biggest problem with childhood and adolescent depression is recognizing the problem and getting the child into treatment," says Michael Sorter, a psychiatrist at Children's Hospital Medical Center in Cincinnati. "The biggest breakthroughs occur when you get the child and family in therapy together with everyone working toward the same goal."[72] The first task, of course, is to recover from the initial depression. After that situation has been brought under control there are a number of "survival tips" that can eliminate or reduce the risk of a recurrence. In the meantime, as psychologist Michael D. Yapko has noted, there is no point in blaming yourself. Depression is not something that anyone chooses.

Still, teens should take an active role in their treatment. It is crucial to remain conscientious about taking prescribed medications. And skipping appointments with a therapist is not a good

A father has a talk with his depressed son. Parents of depressed teens must play an active role in helping their children to recover.

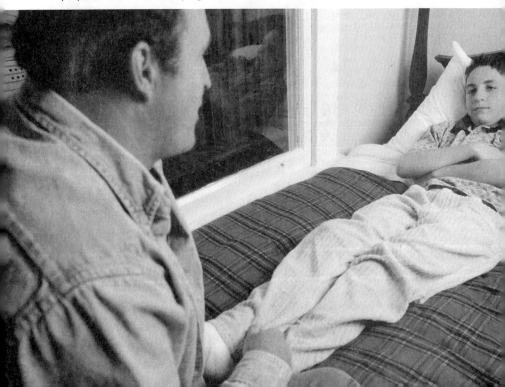

idea—even if everything seems to be going well. Stopping or changing treatment should never be done without first talking it over with one's parents and the professionals who are providing the treatment. Doctors and therapists, after all, can monitor progress, offer encouragement, and adjust medication should problems arise.

One of the reasons some teens find it hard to keep up with treatment is that they do not want to face the fact that depression might return. They would much rather assume that the problem is over once and for all. That is a dangerous attitude, because it makes a recurrence all the more devastating. A much better approach is to remain vigilant for the negative feelings and painful thoughts that might signal a recurrence, while taking responsibility for one's emotional and physical well-being.

A Powerful Ally

Exercise is probably the most overlooked tool in the fight against depression. In fact, it is one of the most powerful and important things anyone can do to prevent depression. Regular exercise releases endorphins, brain chemicals that give people natural highs. Endorphins lift peoples' moods and help them feel energized—the exact opposite of depression. Exercise also helps people sleep more soundly, gives them better appetites, and reduces irritability and anger.

An important study at Duke University followed 156 adults with mild to moderate depression. Those who exercised three times a week did just as well as those who took antidepressants and they were less likely to have recurrences. Another study found the effects of exercise are surprisingly long lasting. Not only did a group of depressed people who exercised three times a week show measurable improvement after only five weeks, but, even if they quit, the benefits lasted for up to a year.

Unfortunately, exercise is usually one of the last things depressed people want to do. Most find it difficult to get started with an exercise program. Yet, even mild exercise a few minutes a day can increase endorphins. Larry was a grossly overweight teen because he ate when he was depressed—and he was depressed almost all the time. Then he discovered the wonders of exercise:

Regular exercise is one of the most effective behaviors teens can adopt to help prevent the onset of depression.

"Exercise, to me, was a well-kept secret. I couldn't believe how good I could feel from working out. Those endorphins are powerful! I'd be at the gym five times a week, for 1 to 2 hours a day. When you're getting rid of one lifetime addiction, sometimes for the first bit you need to replace it with another, healthier addiction."[73]

Most people will need to start slower than Larry did, but the eventual goal should be to do twenty or thirty minutes of continuous exercise several times a week. It is critical to find ways to work out that are enjoyable because it may be too easy to quit otherwise. Exercising with a friend or family member is a good way to keep motivated. No matter what form of exercise is chosen, if it is engaged in regularly a noticeable improvement in mood should result.

Positive Thinking Is Important

Mood is important to monitor because negative thoughts can destroy self-esteem and trigger another depression. One of the reasons that cognitive therapy works so well is that it helps teens realize the debilitating effect of continually "beating oneself up" emotionally. Zoe describes that thought process well: "I never feel enthusiastic about anything. I have very little faith in positive outcomes. I don't believe in myself. I don't trust anyone. I live my life—my job, my dates (if I have any), my friendships—in a fog. I stumble around unable to get very motivated because I'm afraid of what I might run into."[74]

Zoe was amazed at the change in her mood after she was treated for depression. For the first time in her life she realized how different reality looks when negative thoughts are gone. "There's a whole world—a whole way of being—I never knew anything about,"[75] she marveled. Although she will have to continue to monitor her own thoughts, she now knows the power of positive thinking and is much more likely to become engaged with the world.

Friends and Fun

Interaction with other people is another key to preventing depression. Isolation tends to breed depression, while being social helps fight it. Teens who have been isolated for a while may feel

Because isolation can lead to depression, teens who regularly interact with friends reduce their risk of developing the condition.

awkward at first, but renewing friendships or initiating new ones is a great way to combat depression. So is laughter—depression and laughter do not mix well at all. Sharing a joke or a funny situation with others will relieve depressive symptoms and help keep everyone's spirit up.

Finding a Balance

Eating well is another often overlooked way that teens can improve how they feel. Since teens' brains and bodies are developing at a fast pace, good nutrition is critical. Likewise, poor nutrition can be damaging. Unfortunately, in a country where fast foods are readily available, many teen diets are atrocious. Foods that are full of fat

Eating a healthy, low-fat diet rich in vegetables and fruits can help teens ward off depression.

(like cake, doughnuts, french fries, and potato chips) tend to make us feel heavy and sluggish. People who are depressed are probably already feeling tired, so fatty foods will only make them feel worse. While sugary foods like soda and candy provide a brief "sugar high," they can leave teens feeling irritable afterward because their bodies cannot handle the extra sugar in their bloodstream.

Cutting down on sugar and fat (it is not necessary to eliminate them entirely) and eating more fruits, vegetables, and low-fat snacks will help any young person feel healthier and more energized. It is important for teens to be aware of how their bodies reacts to foods, too. For example, some people with depression have discovered that they feel better if they cut out or limit caffeine. A more powerful drug than most people realize, caffeine can disrupt sleep and increase feelings of anxiety. Sunflower seeds and pumpkin seeds, to cite two more examples, contain a natural antidepressant called tryptophan. Snacking on them can help a person remain calm. Eating a balanced diet is an important part of maintaining a person's overall health.

Trusting Ourselves and Others

Finding ways to balance one's emotions is also important. Writing down their thoughts in a journal, diary, or a notebook can help teens get in touch with their true feelings. It worked for Julia: "After a while I was able to recognize habits, conversations, and environments that made me uncomfortable, as well as people who made me feel that way, too. I made notes in my journal about where and when I felt bad. Slowly I began to accept that I could change my circumstances—that I didn't need to see people who hurt me or do things that hurt me."[76]

Becoming involved in art or music is another good way for young people to understand themselves emotionally. Inward-looking activities like these can help ward off negative moods because unexamined feelings are often hidden triggers for depression.

When stress threatens to overwhelm an adolescent, communication with others is also critically important. There are times when every teen needs to talk with someone they can trust to listen and understand. "If you get down, take things as they come,"

When teenagers feel overwhelmed by stress or sadness, communicating with supportive friends is critically important.

advises David, a teen once so crippled by depression that he had to drop out of school. "From my own experience, I know the pain will subside. If you need to talk to someone, you can talk to your friends and family."[77]

If there are no trustworthy parents or friends in a teen's world, he or she may have to seek out a minister, guidance counselor, or other trusted adult. Nineteen-year-old Helene had to look no further than the front of her classroom. "I think that some of my teachers in high school are the reason I'm here today,"[78] she says.

Help can almost always be found if one is willing to look for it. And despite the crippling hopelessness felt by depressed teens, things are almost never as bad as they seem. With effective treatment and vigilance against a recurrence, astounding changes are possible. A fifteen-year-old who still bears the scars of three suicide attempts reflects:

My opinion about myself has changed so much. Nine months ago, I'd look in the mirror and say I hate myself. I wanted to die. Why was I born? I wanted to kill myself. Now I look into the mirror and say what a wonderful person I am. And I'm glad I've been able to climb out of this hole. . . .

Suicide is a permanent solution to a temporary problem. Problems never last forever. Problems are temporary. Problems go away. But life is something very, very precious and worth going through. Live as much as you can. Life is very beautiful. It has its ups and downs, but it's worth it.[79]

Notes

Introduction: The Hopelessness Disease

1. Quoted in Harold S. Koplewicz, *More than Moody: Recognizing and Treating Adolescent Depression.* New York: G.P. Putnam's Sons, 2002, p. 2.
2. Paul C. Quinnett, *Suicide: The Forever Decision.* New York: Crossroad, 2000, p. 45.
3. Quoted in Demitri Papolos and Janice Papolos, *Overcoming Depression.* New York: HarperCollins, 1997.
4. Quoted in Joel Solow, "Teen Depression Strikes Home During the Holidays," *New York Amsterdam News*, December 12, 2002, p. 22.
5. Quoted in preface to Cait Irwin, *Conquering the Beast Within.* New York: Times Books, 1998.
6. Irwin, *Conquering the Beast Within*, p. 8.

Chapter 1: More than Sadness

7. Quoted in Maureen Empfield and Nicholas Bakalar, *Understanding Teenage Depression.* New York: Henry Holt, 2001, p. 17.
8. Quoted in Empfield and Bakalar, *Understanding Teen Depression*, p. 17.
9. Quoted in Empfield and Bakalar, *Understanding Teen Depression*, p. 17.
10. Quoted in Empfield and Bakalar, *Understanding Teen Depression*, p. 17.
11. Irwin, *Conquering the Beast Within*, p. 5.
12. Quoted in Gerald D. Oster and Sarah S. Montgomery, *Helping Your Depressed Teenager.* New York: John Wiley & Sons, 1995, p. 49.
13. Quoted in Joel Solow, "Scars of Adolescent Depression Run

Deep," *New York Amsterdam News*, April 19, 2001, p. 24.
14. Quoted in Charles Cross, *Heavier than Heaven: An Autobiography of Kurt Cobain*. New York: Hyperion, 2001, p. 146.
15. Quoted in Cross, *Heavier than Heaven*, p. 172.
16. Quoted in Harold S. Koplewicz, "More than Moody: Recognizing and Treating Adolescent Depression," *Brown University Child and Adolescent Behavior Letter*, December 2002, p. 1.
17. Quoted in Susan Klenanoff and Ellen Luborsky, *Ups & Downs: How to Beat the Blues and Teen Depression*. New York: Price Stern Sloan, 2001, p. 76.

Chapter 2: Down in the Dumps: Why Teens Get Depressed

18. Quoted in Julia Thorne, *You Are Not Alone: Words of Experience and Hope for the Journey Through Depression*. New York: HarperCollins, 1993, p. 16.
19. Quoted in Kathleen McCoy, *Understanding Your Teenager's Depression*. New York: Berkley, 1994, p. 15.
20. Quoted in Empfield and Bakalar, *Understanding Teen Depression*, p. 17.
21. Quoted in Empfield and Bakalar, *Understanding Teen Depression*, pp. 113–14.
22. Quoted in McCoy, *Understanding Your Teenager's Depression*, p. 41.
23. Quoted in McCoy, *Understanding Your Teenager's Depression*, p. 42.
24. Quoted in McCoy, *Understanding Your Teenager's Depression*, p. 42.
25. Quoted in David Leonhardt, "It's a Girl! (Will the Economy Suffer?)," *New York Times*, October 26, 2003.
26. Koplewicz, *More than Moody*, p. 80.
27. Thorne, *You Are Not Alone*, p. 29.
28. Quoted in McCoy, *Understanding Your Teenager's Depression*, pp. 46–47.
29. Quoted in McCoy, *Understanding Your Teenager's Depression*, p. 54.
30. Quoted in McCoy, *Understanding Your Teenager's Depression*, p. 55.
31. Quoted in Koplewicz, *More than Moody*, p. 218.

Chapter 3: Helpless and Hopeless: The Signs of Depression

32. Quoted in Empfield and Bakalar, *Understanding Teen Depression*, p. 78.
33. Quoted in Beverly Cobain, *When Nothing Matters Anymore: A Survival Guide for Depressed Teens*. Minneapolis, MN: Free Spirit, 1998, pp. 49–50.
34. Quoted in Cobain, *When Nothing Matters Anymore*, p. 50.
35. Quoted in Oster and Montgomery, *Helping Your Depressed Teenager*, p. 47.
36. Quoted in Oster and Montgomery, *Helping Your Depressed Teenager*, p. 42.
37. Koplewicz, *More than Moody*, p. 37.
38. Quoted in Koplewicz, *More than Moody*, pp. 64–65.
39. Quoted in Oster and Montgomery, *Helping Your Depressed Teenager*, p. 47.
40. Koplewicz, *More than Moody*, p. 72.
41. Quoted in Oster and Montgomery, *Helping Your Depressed Teenager*, p. 42.
42. Quoted in Empfield and Bakalar, *Understanding Teen Depression*, p. 108.
43. Koplewicz, *More than Moody*, p. 17.
44. Quoted in Cobain, *When Nothing Matters Anymore*, p. 32.

Chapter 4: When Depression Turns Deadly

45. Quoted in Tad Friend, "Jumpers," *New Yorker*, October 13, 2003, p. 52.
46. Quoted in McCoy, *Understanding Your Teenager's Depression*, p. 56.
47. Quoted in Janice Arenofsky, "Teen Suicide," *Current Health*, December 1997.
48. Cobain, *When Nothing Matters Anymore*, p. 94.
49. Andrew Slaby and Lili Frank Garfinkel, *No One Saw My Pain: Why Teens Kill Themselves*. New York: W.W. Norton, 1994, p. 4.
50. Koplewicz, *More than Moody*, p. 244.
51. Quoted in Koplewicz, *More than Moody*, pp. 249–50.
52. Quoted in Cobain, *When Nothing Matters Anymore*, p. 91.
53. Koplewicz, *More than Moody*, p. 248.
54. Empfield and Bakalar, *Understanding Teen Depression*, p. 62.

55. Quoted in Cobain, *When Nothing Matters Anymore*, p. 103.
56. Irwin, *Conquering the Beast Within*, p. 74.
57. Empfield and Bakalar, *Understanding Teen Depression*, p. 89.

Chapter 5: Restoring Hope: Treating Depression

58. Quoted in Cobain, *When Nothing Matters Anymore*, p. 117.
59. Quoted in Arenofsky, "Teen Suicide."
60. Empfield and Bakalar, *Understanding Teen Depression*, p. 119.
61. Quoted in Oster and Montgomery, *Helping Your Depressed Teenager*, p. 48.
62. Koplewicz, *More than Moody*, p. 37.
63. Oster and Montgomery, *Helping Your Depressed Teenager*, p. 135.
64. Quoted in Empfield and Bakalar, *Understanding Teen Depression*, p. 13.
65. Quoted in Cobain, *When Nothing Matters Anymore*, pp. 140–41.
66. Quoted in Cobain, *When Nothing Matters Anymore*, p. 141.
67. Quoted in Lisa Lee Freeman, "Teen Depression," *Cosmo Girl*, August 2003, p. 103.
68. Quoted in Cobain, *When Nothing Matters Anymore*, p. 121.

Chapter 6: Keeping the Blues at Bay

69. Michael D. Yapko, *Breaking the Patterns of Depression*. New York: Doubleday, 1997, p. 32.
70. Yapko, *Breaking the Patterns of Depression*, p. 32.
71. Laura Epstein Rosen and Xavier Amador, *When Someone You Love Is Depressed: How to Help Your Loved One Without Losing Yourself*. New York: Free Press, 1996, p. 80.
72. Michael Sorter, "The Dark Clouds of Depression," *NEA Today*, December 1994, p. 15.
73. Quoted in Keith G. Kramlinger, ed., *Mayo Clinic on Depression*. Rochester, MN: Mayo Clinic Health Information, 2001, p. 118.
74. Quoted in Thorne, *You Are Not Alone*, p. 64.
75. Quoted in Thorne, *You Are Not Alone*, p. 65.
76. Quoted in Thorne, *You Are Not Alone*, p. 119.
77. Quoted in Cobain, *When Nothing Matters Anymore*, p. 50.
78. Quoted in Arenofsky, "Teen Suicide."
79. Quoted in Solow, "Teen Depression Strikes Home During the Holidays," p. 22.

Organizations to Contact

The American Academy of Child and Adolescent Psychiatry
3615 Wisconsin Ave. NW, Washington, DC 20016-3007
(202) 966-7300
www.aacap.org
This organization assists parents and families in understanding the developmental, behavioral, emotional, and mental disorders affecting children and adolescents.

American Association of Suicidology
4201 Connecticut Ave. NW, Suite 408, Washington, DC 20008
info@suicidology.org
This organization is dedicated to the understanding and prevention of suicide.

Depression and Bipolar Support Alliance (DBSA)
730 N. Franklin St., Suite 501, Chicago, IL 60610
media@dbsalliance.org
The DBSA works to improve the lives of people living with mood disorders. It provides confidential screenings, the latest news regarding depression and bipolar disorders, as well as discussion groups, chat rooms, and information on where to find the nearest support group.

Depression and Related Affective Disorders Association
2330 W. Joppa Rd., Suite 100, Lutherville, MD 21093
drada@jhmi.edu
This organization targets individuals affected by depression, including family members, health care professionals, and the gen-

eral public. Its goal is to alleviate suffering by aiding self-help groups, providing education and information about the latest findings in the field, and supporting research programs. It also works to make people aware of the resources available to persons suffering from depression.

National Foundation for Depressive Illness, Inc.
PO Box 2257, New York, NY 10116
(800) 239-1264
www.health.gov/nhic
The foundation's goals are to educate the public about depressive illness, its consequences and treatment, and to provide information and referrals to anyone who requests help.

For Further Reading

Books

Beverly Cobain, *When Nothing Matters Anymore: A Survival Guide for Depressed Teens*. Minneapolis, MN: Free Spirit, 1998. A practical, well-organized, and readable guide to depression by a cousin of the rock star Kurt Cobain.

Maureen Empfield and Nicholas Bakalar, *Understanding Teenage Depression*. New York: Henry Holt, 2001. A well-written book that explains the dynamics of depression and diagnosis for teens and their parents.

Cait Irwin, *Conquering the Beast Within*. New York: Times Books, 1998. The author recounts her struggle with depression in a book that she illustrated as well. Inspiring, it is one of the few books about teen depression actually written by a teen.

Harold S. Koplewicz, *More than Moody: Recognizing and Treating Adolescent Depression*. New York: G.P. Putnam's Sons, 2002. A useful and hopeful guide to teen depression by a renowned doctor who enjoys working with adolescents. The author does a particularly good job explaining treatment options.

John Preston, *You Can Beat Depression*. San Luis Obispo, CA: Impact, 1996. An easy-to-read, practical guide for overcoming depression. Direct, one-to-one tone of author (who suffered from depression himself) may appeal to teens who might not otherwise read a book on the subject.

Julia Thorne, *You Are Not Alone: Words of Experience and Hope for the Journey Through Depression*. New York: HarperCollins, 1993. Dozens of powerful "in-their-own-words" accounts by people of all ages who have suffered and recovered from depression.

Periodicals

Janice Arenofsky, "Teen Suicide," *Current Health,* December 1997.

Amy Dickinson, "Puppy Love's Bite," *Time,* April 16, 2001.

Lisa Lee Freeman, "Teen Depression," *Cosmo Girl,* August 2003.

Works Consulted

Books

American Medical Association, *Essential Guide to Depression.* New York: Pocket Books, 1998. Filled with pertinent information on depression, portions of this book may be more in depth than necessary for the casual reader.

Charles Cross, *Heavier than Heaven: An Autobiography of Kurt Cobain.* New York: Hyperion, 2001.

Colette Dowling, *You Mean I Don't Have to Feel This Way?* New York: Bantam, 1993. In context of explaining the disease, the author recalls her former husband's bipolar disorder and her daughter's depression. Since it is more than ten years old, book's information on antidepressants may be slightly dated.

Miriam Kaufman, *Overcoming Teen Depression: A Guide for Parents.* Buffalo, NY: Firefly, 2001. A clearly written primer on how teen depression is treated.

Susan Klebanoff and Ellen Luborsky, *Ups & Downs: How to Beat the Blues and Teen Depression.* New York: Price Stern Sloan, 2001. A remarkably complete resource written in an easy-to-read style. The quizzes, cartoons, and firsthand accounts make it very accesible for teens.

Keith G. Kramlinger, ed., *Mayo Clinic on Depression.* Rochester, MN: Mayo Clinic Health Information, 2001. A well-organized and clearly written overview of depression. One of the best single books to help one understand depression.

Kathleen McCoy, *Understanding Your Teenager's Depression.* New York: Berkely, 1994. Offers practical tips for parents along with plenty of examples of what teenage depression looks like.

Works Consulted

Gerald D. Oster and Sarah S. Montgomery, *Helping Your Depressed Teenager*. New York: John Wiley & Sons, 1995. Provides parents with suggestions on how to understand depression. Comprehensive, readable, and concise.

Demitri Papolos and Janice Papolos, *Overcoming Depression*. New York: HarperCollins, 1997. An extremely comprehensive look at the subject but may be too detailed for reluctant readers.

Paul C. Quinnett, *Suicide: The Forever Decision*. New York: Crossroad, 2000. Author approaches the subject with an unusual degree of respect and empathy for the anguish faced by suicidal people.

Laura Epstein Rosen and Xavier Amador, *When Someone You Love Is Depressed: How to Help Your Loved One Without Losing Yourself*. New York: Free Press, 1996. Valuable tips for those who have to live with someone who is depressed.

Andrew Slaby and Lili Frank Garfinkel, *No One Saw My Pain: Why Teens Kill Themselves*. New York: W.W. Norton, 1994. Psychological profiles of teens who either committed suicide or attempted it. Valuable for understanding how suicide can seem like the only option.

Lewis Wolpert, *Malignant Sadness: The Anatomy of Depression*. New York: Free Press, 1999. Author is a biologist inspired to explore the topic after a bout with severe depression. He does a good job explaining what depression feels like "from the inside."

Michael D. Yapko, *Breaking the Patterns of Depression*. New York: Doubleday, 1997. Provides practical information on how to develop the skills to understand and prevent future episodes of depression. Also contains exercises that could be helpful to a depressed teen.

———, *Hand-Me-Down Blues*. New York: Golden Books, 1999. A good look at depression from the family perspective by an insightful clinical psychologist.

Periodicals

Adolescence, "Adolescent Depression and Suicide: A Comprehensive Empirical Intervention for Prevention and Treatment," Spring 2003.

David A. Brent, "Treatment for Adolescent Depression," *Harvard Mental Health Letter*, August 1998.

Brown University Child and Adolescent Behavior Letter, "Adolescent Depression May Lead to Depression Later in Life," July 1994.

Brown University Child and Adolescent Behavior Letter, "Child, Adolescent Depression Distinct from the Adult Version," August 1996.

Betsy Flagler, "Don't Ignore Teen Depression," *Cedar Rapids Gazette*, August 31, 2003.

Tad Friend, "Jumpers," *New Yorker*, October 13, 2003.

"Kids in Pain," *Current Events*, February 2, 2001.

Harold S. Koplewicz, "More than Moody: Recognizing and Treating Adolescent Depression," *Brown University Child and Adolescent Behavior Letter*, December 2002.

David Leonhardt, "It's a Girl! (Will the Economy Suffer?)", *New York Times*, October 26, 2003.

Mark Nichols, "The Quest for a Cure," *Maclean's*, December 1, 1997.

Gilles Pinette, "What Is Teen Depression," *Raven's Eye*, September 2001.

Joel Solow, "Scars of Adolescent Depression Run Deep," *New York Amsterdam News*, April 19, 2001.

———, "Teen Depression May Require Alternative Therapy," *Brown University Child and Adolescent Behavior Letter*, November 1994.

———, "Teen Depression Strikes Home During the Holidays," *New York Amsterdam News*, December 12, 2002.

Michael Sorter, "The Dark Clouds of Depression," *NEA Today*, December 1994.

Women's Health Weekly, "Rumination Tied to Gender Differences in Adolescent Depression Rates," August 17, 1998.

Index

abuse, drug/alcohol. *See* substance abuse
abuse, emotional/physical, 21, 23, 29
academic performance
  college and, 31–32
  grades and, 17
  learning disorders and, 29
  parents' expectations and, 21, 26–27
alcohol. *See* substance abuse
Amador, Xavier, 71
American College Health Association, 32
anger, 36–37
Angold, Adrian, 69
antidepressants
  brain chemicals and, 63–64
  discovery of, 69
  dosage adjustments of, 64–65, 67
  function of, 63–64
  side effects of, 64
  therapy and, 65–66
  see also Prozac
anxiety, 23, 32, 35
appetite. *See* eating disorders

behavioral disorders, 12–13, 29, 36–37
bipolar disorder, 45–46
blood tests, 69
brain function
  changes in, 12–13, 67, 69
  diagram of, 68
  endorphins and, 73–74
  gray matter and, 67, 69

causes
  abuse, 21, 23, 29
  biological influences, 69–71
  college transition, 31–32
  family, 21, 24–26
  heredity, 18, 23
  loss, 29–30
  peer pressure, 21–22, 35
  see also academic performance
Cavett, Dick, 7–8
Centers for Disease Control, 50
chemical imbalance, 18, 34–35
Children's Hospital Medical Center (Cincinnati), 72
Cobain, Beverly, 48–49
Cobain, Kurt, 16–17, 48, 50
Conquering the Beast Within (Irwin), 15
counseling, 58

depression
  adults and, 18, 33, 42
  diagnosis of, 10, 13, 19
    misdiagnosis and, 42–43
    underdiagnosis and, 17
  as a disease, 10–11, 70–71
  duration of, 12, 19
  girls and, 27–29, 50
  as a mental disorder, 11–12
  risk of future episodes of, 10
  syndrome for, 12
  varieties of, 43–46
  see also causes; prevention; symptoms; treatment
Division of Adolescent Medicine at

Children's Hospital (Los
    Angeles), 21
divorce, 24–26, 28
dopamine, 34
drug abuse. See substance abuse
Duke University, 27, 73
dysthymic disorder, 44

eating disorders, 11–13, 15, 32–33
Empfield, Maureen, 53, 56, 58
employment, 21
exercise, 73–74

families, dysfunctional, 25–26
foster care, 24
Frederick, Calvin, 47

Gardner, Lee Robbins, 31
gay teens, 29
genetics, 23–24, 69, 71
Giedd, Jay, 67
grades. See academic performance
guilt, 14–15, 22, 37

Helping Your Depressed Teenager
    (Oster), 23
heredity, 18, 23
hopelessness, 15–17, 47–48
hormones, 38, 42, 69
hospitalization, 55–56, 66–67

Imrie, Marissa, 47
Irwin, Cait, 9, 15, 56

Katz, Steven, 26
Koplewicz, Harold S., 17, 29, 37,
    40, 50, 60

learning disorders. See academic
    performance
limbic system, 35

MacKenzie, Richard, 21
major depressive disorder (MDD),
    6–9, 41, 44–45, 66
manic depression, 46

mood swings, 12, 38
Moretti, Enrico, 28

National Institute of Mental
    Health, 67
neurotic depression. See dysthymic
    disorder
neurotransmitters, 34
New York University Child Center,
    17, 37
Nirvana, 16
norepinephrine, 34
Northern Westchester Medical
    Hospital, 53
Northwestern University
    Memorial Hospital, 26
nutrition, 76–77
    see also eating disorders

Oster, Gerald D., 23

pain, physical, 15, 18
peer pressure, 27, 30–32, 35
personality disturbances, 32
pregnancy, 17
prevention
    communication and, 77–78
    early diagnosis and, 19, 72
    emotional balance and, 77
    exercise and, 73
    interaction with people and,
        75–76
    nutrition and, 76–77
    positive thinking and, 74–75
    regularity of treatment and,
        72–73
Prozac, 63, 65
psychiatrists, 55–56, 66
psychoanalytic techniques, 61
psychosis, 53
psychotherapy, 58
puberty, 28–30, 69

Quinnett, Paul C., 7

reactive depression, 44

Index

relaxation training, 69
Rosen, Laura Epstein, 71

self-mutilation, 41
September 11, 2001, 70
serotonin, 34, 63, 69
severe depression. *See* major depressive disorder
sex
  gay teens and, 29
  harassment of girls and, 28
  as an outlet, 17
  puberty and, 28–30, 69
shame, 14–15, 37
Slaby, Andrew, 49
sleep disturbances, 12–13, 15, 33, 41
social pressures, 17–18, 29, 40
Sorter, Michael, 72
spinal taps, 69
statistics/surveys
  on adults who suffered depression as teens, 58
  on college students feeling intense hopelessness, 32
  on girls with divorced parents, 28
  on mothers with depression, 24
  on peak ages for depression, 18, 24
  on rates of depression, 6, 11, 18
  on substance abuse, 40–41
  on suicide, 17, 50, 69
stress, 70–71, 77
substance abuse, 17, 23, 32, 40–42, 53
suicide
  hopelessness/worthlessness and, 15–17, 47–48
  hospitalization and, 55–56, 66–67
  methods of, 50
  prevention of, 53–56
  selfishness of, 14–15
  substance abuse and, 53
  as symptom of depression, 12
  talk of, 41–42
  warning signs of, 51–53
  see also statistics/surveys, on suicide
Suicide: The Forever Decision
(Quinnett), 7
symptoms
  eating disorders, 11–13, 15, 32–33
  inability to concentrate, 40
  lack of interest, 38–39
  length of, 38
  low self-esteem, 39–40
  recognition of, 37–38
  self-destructive behavior, 41
  sleep disturbances, 12–13, 15, 33, 41
  suicide, 12
synaptic transmission, 33–34

tests, 69
therapy
  cognitive, 59–61, 69, 74–75
  family sessions and, 61–62
  group, 62, 69
  interpersonal (IPT), 61–62
  ongoing, 63
  psychiatric treatment and, 56
  talk, 57–62, 66
  see also antidepressants
Thorne, Julia, 29–30
treatment
  exercise and, 73–74
  hospitalization and, 55–56, 66–67
  medications and, 19, 35, 56–60, 72
  regularity of, 72–73
  residential programs and, 36
  success of, 19
  tests and, 69
  understanding depression and, 70–72
  see also therapy

University of California, 28
University of Oregon, 11
U.S. Surgeon General, 6

World Health Organization, 11
World Trade Center, 70

Yapko, Michael D., 70, 72

Picture Credits

Cover image: © Corel Corporation
© Annebicque Bernard/CORBIS SYGMA, 65
© Peter Byron/Photo Edit, 48
© Philip James Corwin/CORBIS, 54
© Mary Kate Denny/Photo Edit, 49
© Richard Hutchings/CORBIS, 45
© Richard Hutchings/Photo Edit, 28
The Image Bank/Getty Images, 43
© Michael Keller/CORBIS, 76
© Michael Newman/Photo Edit, 55
© Gabe Palmer/CORBIS, 27
PhotoDisc, 7, 11, 13, 18, 21, 22, 25, 31, 36, 40, 51, 72, 74, 75, 78
Photos.com, 71
Reuters/Lee Celano/Landov, 16
© Chuck Savage/CORBIS, 30
© Ariel Skelley/CORBIS, 14
© Tom Stewart/CORBIS, 66
Stone/Getty Images, 52, 62
Taxi/Getty Images, 60
Jim Varney/Photo Researchers, Inc., 59
© David Young-Wolff/Photo Edit, 39
Steve Zmina, 68

About the Author

A former editor at *Reminisce* magazine, Michael J. Martin is a freelance writer whose home overlooks the Mississippi River in Lansing, Iowa. He has written more than a dozen books for young people, as well as magazine articles for publications like *Boys' Life* and *Timeline*, and has a master's degree in educational psychology from the University of Wisconsin–Milwaukee. His most recent books for Lucent are a biography of test pilot Chuck Yeager and *The Korean War: Life as a POW*.